I0049483

Materials Processing and Crystal Growth for Thermoelectrics

Materials Processing and Crystal Growth for Thermoelectrics

Special Issue Editor

George S. Nolas

MDPI • Basel • Beijing • Wuhan • Barcelona • Belgrade

MDPI

Special Issue Editor
George S. Nolas
University of South Florida
USA

Editorial Office
MDPI
St. Alban-Anlage 66
4052 Basel, Switzerland

This is a reprint of articles from the Special Issue published online in the open access journal *Catalysts* (ISSN 2073-4344) from 2017 to 2018 (available at: https://www.mdpi.com/journal/crystals/special_issues/thermoelectrics).

For citation purposes, cite each article independently as indicated on the article page online and as indicated below:

LastName, A.A.; LastName, B.B.; LastName, C.C. Article Title. *Journal Name* **Year**, *Article Number*, Page Range.

ISBN 978-3-03897-588-5 (Pbk)
ISBN 978-3-03897-589-2 (PDF)

© 2019 by the authors. Articles in this book are Open Access and distributed under the Creative Commons Attribution (CC BY) license, which allows users to download, copy and build upon published articles, as long as the author and publisher are properly credited, which ensures maximum dissemination and a wider impact of our publications.

The book as a whole is distributed by MDPI under the terms and conditions of the Creative Commons license CC BY-NC-ND.

Contents

About the Special Issue Editor

George S. Nolas, Distinguished University Professor, University of South Florida. Prof. Nolas is a Fellow of the American Association for the Advancement of Science and the American Physical Society. Prof. Nolas' expertise is in the area of condensed matter physics and materials science, including materials for thermoelectrics power generation and refrigeration applications. Prof. Nolas holds several patents, has published over 200 peer-reviewed journal articles, several book chapters, and two books including the foremost text in the field of thermoelectrics. Prof. Nolas has also been honored with four teaching and mentorship awards, and his students have been recognized by dissertation awards, research scholarships, and fellowships.

Preface to "Materials Processing and Crystal Growth for Thermoelectrics"

A growing public awareness has resulted in consensus that new technologies for renewable energy must be realized in the near future. This has lead to a focus on several different solutions to this problem. Thermoelectrics can play a role in this regard, and is one technology that continues to be of interest. Thermoelectric devices are especially attractive since they have no moving parts, are very reliable, and allow for a wide range of applications, from industrial to consumer applications. In order to efficiently convert energy using thermoelectricity, certain material properties are desirable. This includes a high electrical conductivity, σ, to maintain high charge current, a high Seebeck coefficient, S, to maintain a high voltage drop, and a low thermal conductivity, κ, to maintain the temperature gradient. The performance of a thermoelectric device is characterized by the figure of merit, a dimensionless parameter defined as $ZT = S^2\sigma/\kappa$, where T is the absolute temperature. All other aspects being equal, materials with larger ZT values result in more efficient thermoelectric devices. New materials research is therefore essential. It is our hope that the manuscripts contained in this volume will provide a concise reference to some of the current research in the field of thermoelectric materials research.

George S. Nolas
Special Issue Editor

crystals

MDPI

Article

Simultaneous Enhancement of Electrical Conductivity and Seebeck Coefficient of [6,6]-Phenyl-C71 Butyric Acid Methyl Ester (PC$_{70}$BM) by Adding Co-Solvents

Mina Rastegaralam [1], Changhee Lee [1] and Urszula Dettlaff-Weglikowska [2,*]

[1] Department of Electrical and Computer Engineering, Inter-University Semiconductor Research Center, Seoul National University, 1 Gwanak-ro, Gwanak-gu, Seoul 08826, Korea; mn.rstgrlm@gmail.com (M.R.); chlee7@snu.ac.kr (C.L.)
[2] Materials Science Consulting and Management, 22113 Oststeinbek, Germany
* Correspondence: udettlaff.w@gmail.com

Received: 8 April 2018; Accepted: 22 May 2018; Published: 26 May 2018

Abstract: Chemical modification by co-solvents added to [6,6]-Phenyl-C71 butyric acid methyl ester, commonly known as an n-type semiconducting fullerene derivative PC$_{70}$BM, is reported to change the electrical and thermoelectric properties of this system. Power factor of the casted PC$_{70}$BM samples achieves values higher than that determined for a variety of organic compounds, including conducting polymers, such as PEDOT:PSS in the pristine form. After chemical functionalization by different solvents, namely N,N-Dimethylformamide (DMF), dimethyl sulfoxide (DMSO), N-Methyl-2-pyrrolidone (NMP), acetonitrile (AC), and 1,2-Dichloroethane (DCE), the four-probe in-plane electrical conductivity and Seebeck coefficient measurements indicate a simultaneous increase of the electrical conductivity and the Seebeck coefficient. The observed effect is more pronounced for solvents with a high boiling point, such as N,N-Dimethylformamide (DMF), dimethyl sulfoxide (DMSO), and N-Methyl-2-pyrrolidone (NMP), than in acetonitrile (AC) and 1,2-Dichloroethane (DCE). We identified the origin of these changes using Hall mobility measurements, which demonstrate enhancement of the PC$_{70}$BM charge carrier mobility upon addition of the corresponding solvents due to the improved packaging of the fullerene compound and chemical interaction with entrapped solvent molecules within the layers.

Keywords: electrical conductivity; Seebeck coefficient; power factor; PC$_{70}$BM; figure of merit

---

1. Introduction

Thermoelectric materials are very effective at turning a temperature difference directly into electricity. These materials can contribute to both cooling and thermoelectric power generation [1,2]. Traditionally, inorganic materials, such as Bi$_2$Sb$_3$, Bi$_2$Te$_3$, and PbTe, have been used for thermoelectric applications [3]. The problems with using these inorganic compounds are high production costs, toxicity, and scarcity of materials [3,4]. To overcome these problems, organic thermoelectric materials have attracted considerable attention due to their advantages, such as non-toxicity, low cost, mechanical flexibility, abundant raw materials, solution processability, and low thermal conductivity [3–7]. To evaluate the efficiency of a thermoelectric system, a dimensionless quantity called thermoelectric figure of merit ZT is applied. Figure of merit is defined as ZT = S^{2}σT/κ, where S is the Seebeck coefficient, σ is the electrical conductivity, T is the absolute temperature, and κ is the thermal conductivity. Thermoelectric materials with high efficiency have a high Seebeck coefficient, high electrical conductivity, low thermal conductivity, and therefore high ZT. To design an effective thermoelectric material, it is necessary to increase electrical conductivity and Seebeck coefficient simultaneously, while keeping κ constant. However, this is a challenging task as an increase in the number of carriers from doping will sacrifice Seebeck coefficient. One way towards improving ZT is to make use of chemical functionalization that increases mobility in the material, maintaining a constant number of carriers, which in turn leads to improving both electrical conductivity and Seebeck coefficient, according to the equation $\sigma = en\mu$, where e is the electron charge, n is the charge carrier density, and μ is the carrier mobility [8].

To date, the thermoelectric properties of a variety of organics, such as the following conducting polymers, have been studied: poly [3-hexylthiophene] (P3HT), poly [*N*-90-heptadecanyl-2,7-carbazole-alt-5,5-(40,70-di-2-thienyl-20,10,3-benzothiadizole)] (PCDTBT), polyacetylenes, polyaniline, polypyrrole, poly(paraphenylene), poly(p-phenylenevinylene), poly(carbazolenevinylene) derivative, poly(3,4-ethylenedioxythiophene) polystyrene sulfonate (PEDOT/PSS), FBDPPV, and Poly({4,8-bis [(2-ethylhexyl)oxy] benzo [1,2-*b*:4,5-*b'*] dithiophene-2,6-diyl}{3-fluoro-2-[(2-ethylhexyl) carbonyl] thieno[3,4-*b*] thiophenediyl}), known as PTB7 [9–17].

$PC_{70}BM$ is a fullerene derivative compound showing electron-transporting properties and a potential for a variety of applications in polymer solar cells and organic electronics [18,19]. However, its low electron mobility and low electrical conductivity are limiting factors for application in practical devices. An effective way to increase the electrical conductivity of fullerenes is chemical doping. Compared with the progress made on p-doping, n-doping is lagging due to the difficulties in finding efficient and stable dopants for n-type organic semiconductors.

Here, we report on thermoelectric properties of pristine $PC_{70}BM$ and simultaneous enhancement of its electrical conductivity and Seebeck coefficient, and demonstrate a proof of principle for material modification through addition of co-solvents. The following section contains experimental details related to preparation of the samples, determination of their electrical conductivity, Seebeck coefficients, and charge carrier mobility, followed by a discussion of achieved results.

2. Experimental

Figure 1 shows the molecular structure of the applied chemical materials and Table 1 shows the boiling points of the solvents. PCBM was purchased from 1-Material. Chlorobenzene (CB), *N,N*-Dimethylformamide (DMF), Dimethyl Sulfoxide (DMSO), N-Methyl-2-pyrrolidone (NMP), Acetonitrile (AC), and 1,2-Dichloroethane (DCE) were all purchased from Sigma Aldrich (Yongin, Kyungi, Korea) and used as received. Six solutions were prepared by dissolving 40 mg of $PC_{70}BM$ in 1 mL of chlorobenzene with and without adding 0.1 mL of DMF, DMSO, NMP, DCE, and AC under an argon atmosphere of a glove box. Samples were fabricated by drop casting of the solutions on glass substrates, which were cleaned by sonication in acetone, isopropyl alcohol, and water, and treated with UV ozone for 15 min before use. The cast films were dried for 24 h in a glove box. The thicknesses of the samples were measured by the profilometer and were found to be 3.04 ± 0.06 μm, on average. The in-plane electrical conductivity was measured for samples by the standard four-probe Van der Pauw method at room temperature. The contacts were placed at the corners of each sample. For comparison the out of plane conductivity of the pristine sample with 1 μm thickness was also measured using an electron-only device with the structure shown in Figure 2. LiF and Al electrodes (deposited through a shadow mask) were thermally evaporated in a ~10^{-6} Torr vacuum with a 0.5 nm and 100 nm thickness, respectively. The Seebeck coefficient was calculated using $S = -\Delta V/\Delta T$, where ΔV was the thermoelectric voltage generated along the sample when it was subjected to the temperature difference, ΔT, as shown schematically in Figure 3. Two K-type thermocouples were used to measure the temperature on both ends of the samples.

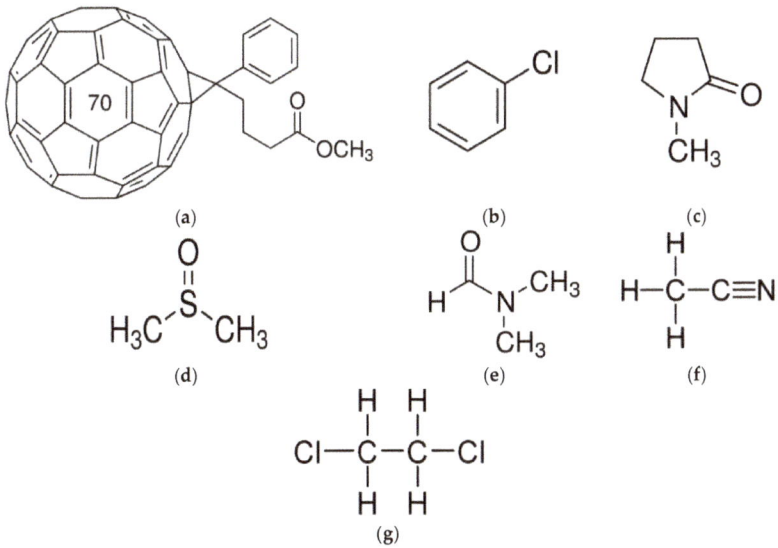

Figure 1. Molecular structure of the applied chemical materials. (**a**) PCBM; (**b**) CB; (**c**) NMP; (**d**) DMSO; (**e**) DMF; (**f**) AC; (**g**) DCE.

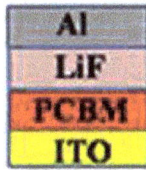

Figure 2. Electron-only device structure.

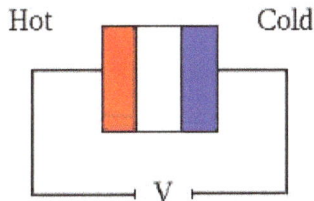

Figure 3. Schematic presentation of the Seebeck coefficient determination.

Table 1. Boiling points of the solvents.

Material	Boiling Point (°C)
Chlorobenzene	131
1-Methyl-2-pyrrolidone	202
Dimethylsulfoxide	189
N,N-Dimethylformamide	153
1,2-Dichloroethane	83
Acetonitrile	82

3. Results and Discussion

$PC_{70}BM$ can be chemically modified by a variety of chemicals that affect electrical conductivity by doping or de-doping. To find out the effect of co-solvents on the thermoelectric properties of the $PC_{70}BM$ samples, a series of solvents with different chemical structures, boiling points, and polarities were selected to be added to the $PC_{70}BM$ solution in chlorobenzene for manufacturing samples. Figure 4 reveals the results of in-plane electrical conductivity measurements of the $PC_{70}BM$ samples. The diagram shows that the electrical conductivity has increased upon addition of the second solvents. In particular, addition of solvents with higher boiling points leads to larger enhancement of the electrical conductivity by a factor of 3. Considering the boiling point of the solvents, it is clear that the samples in which the solvents with higher boiling points were used need a longer time for drying to complete evaporation. This low evaporation rate allows for better ordering within the molecular packing of the fullerene derivative $PC_{70}BM$, and therefore, its ability to transport charge carriers improves, leading to increased electrical conductivity. The best performance was obtained after addition of DMSO. The electrical conductivity values of the samples are 0.108 S/m, 0.135 S/m, 0.156 S/m, 0.25 S/m, 0.307 S/m, and 0.323 S/m for pristine, CB:DCE, CB:AC, CB:DMF, CB:NMP, and CB:DMSO samples, respectively. We anticipated that electrical conductivity of the $PC_{70}BM$ will be anisotropic depending on whether the measurement was performed along the layer or perpendicular to it, because the lateral electrical transport depends on the arrangement of fullerene molecules on the plane, while the perpendicular electrical transport is determined by the stacking of individual layers. Indeed, for the $PC_{70}BM$ sample, we observed strong anisotropy of conductivity. The out of plane conductivity of the pristine sample was found to be 2.6×10^{-4} S/m, which is 3 orders of magnitude lower than that of in-plane conductivity. However, for the thermoelectric applications, the in-plane conductivity is essential, as a layered structure is preferential for the multiple p-n junctions connected in series in practical devices.

The effect of chemical modification of $PC_{70}BM$ was further investigated by way of measuring the charge carrier mobility in the prepared series of samples. Figure 5 shows that the enhancement of conductivity is accompanied by the increase of the mobility upon adding of the second solvent.

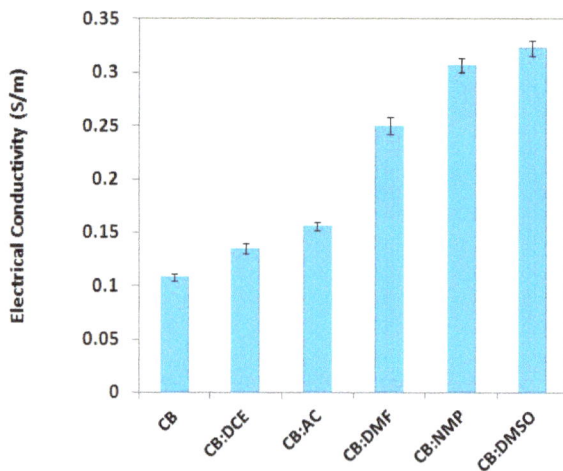

Figure 4. Electrical conductivity of the samples.

The nature of charge carriers in the pristine and treated PCBM samples was determined by the thermoelectric power measurement. The corresponding Seebeck coefficients of the samples are shown in Figure 6. The negative sign of the Seebeck coefficient indicates that the major charge carriers are electrons, confirming that all $PC_{70}BM$ samples are n-type semiconductor, in pristine form and after treatment by addition of the second solvent. The Seebeck coefficient values of the samples are -433 μV/K, -440 μV/K, -440 μV/K, -448 μV/K, -461 μV/K, and -464 μV/K for pristine, CB:DCE, CB:AC, CB:DMF, CB:NMP, and CB:DMSO samples, respectively.

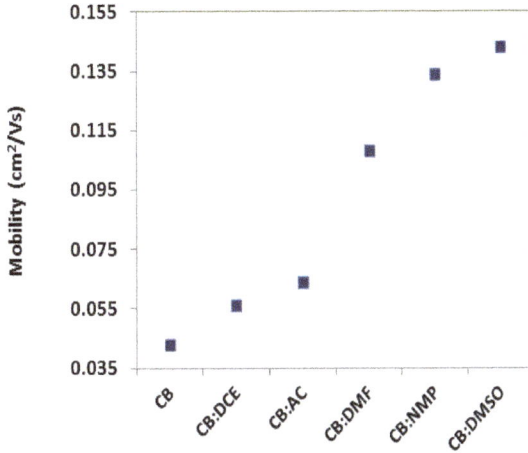

Figure 5. Charge carrier mobility of the samples.

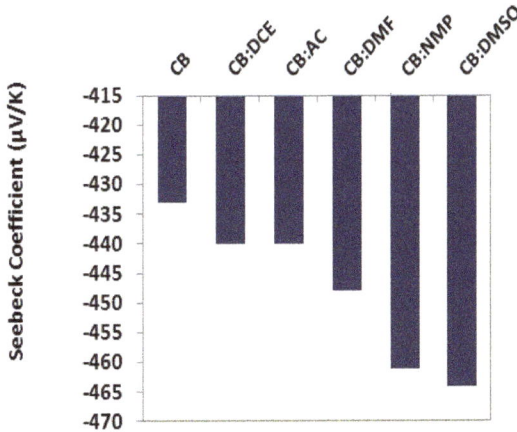

Figure 6. Seebeck coefficient values of the samples.

Previously, it was shown that electrical conductivity increases with increasing boiling point of the used solvent. Similarly, Figure 6 shows that the absolute values of the Seebeck coefficients increase with the increasing boiling point of the solvent as well. The highest Seebeck coefficient was obtained from the CB:DMSO system. Although the applied solvents containing heteroatoms can push or extract electrons from the conducting system, the chemical doping has not been identified within the predefined experimental conditions. Hence, the effect of applying chemical compounds with high boiling points on the thermopower of the $PC_{70}BM$ can be explained as follows. While electrical conductivity is associated with the transport of all mobile charges (holes and electrons), the Seebeck coefficient is related to the transport of energetic charges. Transport of charge carriers within the samples under a temperature gradient is described by the heterogeneous model involving highly conductive regions separated by barrier regions, such as fullerene inter-junctions and fullerene-solvent interfaces, by a hopping mechanism [13]. Increasing the number of nano-scale barriers in the form of interfaces in heterogeneous materials, such as the investigated arrangement of fullerene derivative network, interacting with the trapped chemical molecules of intentionally added solvents within the film is expected to enhance the thermopower. Therefore, we observed an enhancement of Seebeck coefficient for samples including DCE, AC, DMF, NMP, and DMSO.

The increase of Seebeck coefficient and conductivity in parallel in a hole-conducting system has been demonstrated previously for some doping levels of intentionally doped poly(alkylthiophene), in which ground state hole carriers, created by doping with a minor additive component, were mainly at an orbital energy set below the hole energy of the major component of a blend [9]. Similarly, the observed phenomena in our system can be explained using the electronic band structure of $PC_{70}BM$ and the position of the Fermi level in the density of states of the composition consisting of $PC_{70}BM$ and trapped solvent molecules. Energy levels of such molecular semiconductors are determined primarily by molecular orbitals, and their energy can be derived electrochemically, spectroscopically, or by means of semi-empirical calculations. Gao et al. have proposed a numerical model applicable to doped organic materials showing that the Seebeck coefficient increases as greater proportions of the electrical conductivity occur at energies different from the Fermi level [20]. This depends on the existence of states away from the Fermi level, the probability of their occupancy, and the charge carrier mobility in these states. Wuesten et al. extended the model used by Gao et al. allowing hopping transport and the effects of defects typical for organics, such as grain boundaries, that produce additional localized states [21]. Both models predict the experimental behavior of a variety of organic materials showing that high doping levels decrease S because the Fermi level is brought closer to the energy level where charge transport is favored. The above-mentioned models can be applied to our preliminary results regarding the establishment of a Fermi level with respect to the orbital energies of the two molecular components. In our system of two components, the conductive molecular semiconductor $PC_{70}BM$ as a main component has carrier energies just above the orbital energies of the trapped molecules of the added solvent that form the minor component. The selected solvents exhibit different molecular structures, polarity, and redox potential with respect to $PC_{70}BM$. The Fermi level is established by the additive, while the current from injected charge is carried predominantly in the higher energy orbitals of the bulk composition. Hence, thermal excitation of additive-generated charge carriers in one region of the sample can lead to the migration of some of them into the bulk energy levels of another region of the sample, increasing thermoelectric voltage. On the other hand, the Seebeck coefficient increases in the presence of a strong phonon-electron coupling and counteracts the effects of the Seebeck coefficient decrease caused by enhanced charge carrier mobility [22,23]. However, these hypotheses are only speculations and require further study.

To find out how the mixed solvents affect the morphology of the $PC_{70}BM$ samples, we investigated the sample surface by using an atomic force microscope (AFM). Figure 7 shows the corresponding AFM images. The sample morphology of the prepared $PC_{70}BM$ films does not significantly change. However, the surface roughness slightly decreases when the second solvent is added, indicating that more compact and ordered structures are formed, which could imply higher charge carrier mobility.

The root-mean-square (RMS) surface roughness was found to be 0.72, 0.62, 0.45, 0.37, 0.35, and 0.3 nm for pristine, CB:DCE, CB:AC, CB:DMF, CB:NMP and CB:DMSO samples, respectively.

Figure 8 shows the thermoelectric power factor of our samples calculated using $P = S^2\sigma$. The highest power factor was obtained from the DMSO sample, which shows that among these solvents, DMSO is the best one for preparing PCBM solutions to obtain samples with a higher thermoelectric power factor.

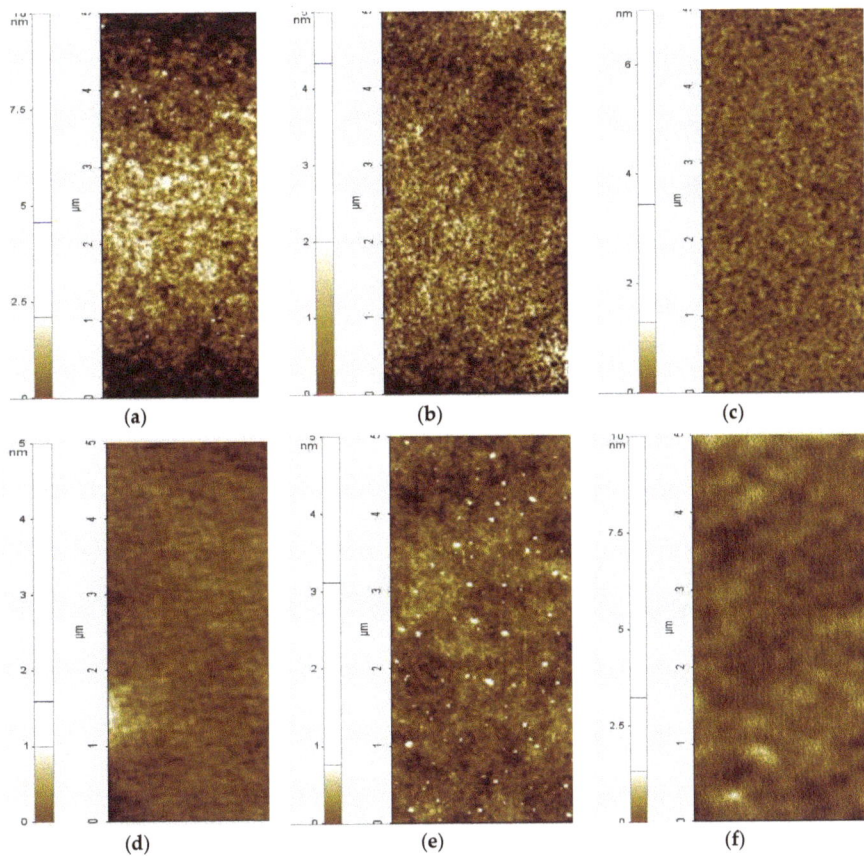

Figure 7. AFM images of the samples. (**a**) CB; (**b**) CB:DCE; (**c**) CB:AC; (**d**) CB:DMF; (**e**) CB:NMP; (**f**) CB:DMSO.

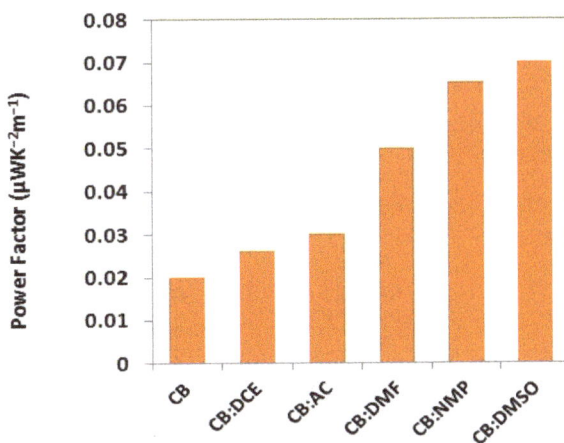

Figure 8. Thermoelectric power factor values of pristine and treated PCBM.

The calculated values of the power factor are 0.02, 0.026, 0.030, 0.05, 0.065, and 0.07 $\mu WK^{-2}m^{-1}$ for pristine, CB:DCE, CB:AC, CB:DMF, CB:NMP, and CB:DMSO, respectively. It is worth noting that the power factor value of $PC_{70}BM$ is higher than a variety of conducting polymers, including poly(3,4-ethylenedioxythiophene) polystyrene sulfonate (PEDOT/PSS), poly [*N*-90-heptadecanyl-2,7-carbazole-alt-5,5-(40,70-di-2-thienyl-20,10,3-benzothiadizole)] (PCDTBT), poly(paraphenylene), poly(p-phenylenevinylene), poly(carbazolenevinylene) derivative, and polyaniline [10,11,14,15].

4. Conclusions

Using the procedure of mixed-solvents for the $PC_{70}BM$, we have demonstrated that both thermoelectric parameters, electrical conductivity, and the Seebeck coefficient can be manipulated in such a way that a simultaneous enhancement of both factors can be achieved. Due to the entrapped molecules of solvents in the PCBM network, an increase in the charge carrier mobility and the Seebeck coefficient was possible during maintaining a constant number of charge carriers. The simultaneous enhancement in electrical conductivity and the Seebeck coefficient closely correlates with the experimentally determined charge carrier mobility. From a series of investigated solvents, DMSO and NMP with higher boiling points appeared to be the most effective co-solvents in improving the power factor.

Author Contributions: M.R. conceived, designed, and performed the experiments; U.D.-W. coordinated scientific work and interpreted the experimental results; and C.L. wrote the paper. All authors read and approved the final version of the manuscript to be submitted.

Acknowledgments: This work was supported by the National Research Foundation of Korea (NRF-2016R1A2B3009301) and LG Chem, Ltd., Korea.

Conflicts of Interest: The authors declare no conflicts of interest.

References

1. Poudel, B.; Hao, Q.; Ma, Y.; Lan, Y.; Minnich, A.; Yu, B.; Yan, X.; Wang, D.; Muto, A.; Vashaee, D.; et al. High-Thermoelectric Performance of Nanostructured Bismuth Antimony Telluride Bulk Alloys. *Science* **2008**, *320*, 634–638. [CrossRef] [PubMed]
2. Zhao, W.; Fan, S.; Xiao, N.; Liu, D.; Tay, Y.Y.; Yu, C.; Sim, D.; Hng, H.H.; Zhang, Q.; Boey, F.; et al. Flexible Carbon Nanotube Papers with Improved Thermoelectric Properties. *Energy Environ. Sci.* **2012**, *5*, 5364–5369. [CrossRef]
3. Culebras, M.; Gómez, C.M.; Cantarero, A. Thermoelectric measurements of PEDOT:PSS/expanded graphite composites. *J. Mater. Sci.* **2013**, *48*, 2855–2860. [CrossRef]
4. Dubey, N.; Leclerc, M. Conducting Polymers: Efficient Thermoelectric Materials. *J. Polym. Sci. B Polym. Phys.* **2011**, *49*, 467–475. [CrossRef]
5. Bubnova, O.; Khan, Z.U.; Malti, A.; Braun, S.; Fahlman, M.; Berggren, M.; Crispin, X. Optimization of the thermoelectric figure of merit in the conducting polymer poly(3,4-ethylenedioxythiophene). *Nat. Mater.* **2011**, *10*, 429–433. [CrossRef] [PubMed]
6. Zhang, Q.; Sun, Y.M.; Xu, W.; Zhu, D.B. Thermoelectric energy from flexible P3HT films doped with a ferric salt of triflimide anions. *Energy Environ. Sci.* **2012**, *5*, 9639–9644. [CrossRef]
7. Gao, F.; Liu, Y.; Xiong, Y.; Wu, P.; Hu, B.; Xu, L. Fabricate organic thermoelectric modules use modified PCBM and PEDOT:PSS materials. *Front. Optoelectron.* **2017**, *10*, 117–123. [CrossRef]
8. Petsagkourakis, I.; Pavlopoulou, E.; Portale, G.; Kuropatwa, B.A.; Dilhaire, S.; Fleury, G.; Hadziioannou, G. Structurally-driven enhancement of thermoelectric properties within poly(3,4-ethylenedioxythiophene) thin films. *Sci. Rep.* **2016**, *6*, 30501. [CrossRef] [PubMed]
9. Sun, J.; Yeh, M.L.; Jung, B.J.; Zhang, B.; Feser, J.; Majumdar, A.; Katz, H.E. Simultaneous increase in Seebeck coefficient and conductivity in a doped poly(alkylthiophene) blend with defined density of states. *Macromolecules* **2010**, *43*, 2897–2903. [CrossRef]

10. Maiz, J.; Rojo, M.M.; Abad, B.; Wilson, A.A.; Nogales, A.; Borca-Tasciuc, D.A.; Borca-Tasciuc, T.; Martín-González, M. Enhancement of thermoelectric efficiency of doped PCDTBT polymer films. *RSC Adv.* **2015**, *5*, 66687–66694. [CrossRef]

11. Xuan, Y.; Liu, X.; Desbief, S.; Leclère, P.; Fahlman, M.; Lazzaroni, R.; Berggren, M.; Cornil, J.; Emin, D.; Crispin, X. Thermoelectric properties of conducting polymers: The case of poly(3-hexylthiophene). *Phys. Rev. B* **2010**, *82*, 115454. [CrossRef]

12. Mateeva, N.; Niculescu, H.; Schlenoff, J.; Testardi, L.R. Correlation of Seebeck coefficient and electric conductivity in polyaniline and polypyrrole. *J. Appl. Phys.* **1998**, *83*, 3111–3117. [CrossRef]

13. Kaiser, A.B. Thermoelectric power and conductivity of heterogeneous conducting polymers. *Phys. Rev. B* **1989**, *40*, 2806. [CrossRef]

14. Toshima, N. Conductive polymers as a new type of thermoelectric material. *Macromol. Symp.* **2002**, *186*, 81–86. [CrossRef]

15. Jiang, Q.; Liu, C.; Song, H.; Shi, H.; Yao, Y.; Xu, J.; Zhang, G.; Lu, B. Improved thermoelectric performance of PEDOT:PSS films prepared by polar-solvent vapor annealing method. *J. Mater. Sci. Mater. Electron.* **2013**, *24*, 4240–4246. [CrossRef]

16. Ma, W.; Shi, K.; Wu, Y.; Lu, Z.Y.; Liu, H.Y.; Wang, J.Y.; Pei, J. Enhanced molecular packing of a conjugated polymer with high organic thermoelectric power factor. *ACS Appl. Mater. Interfaces* **2016**, *8*, 24737–24743. [CrossRef] [PubMed]

17. Rastegaralam, M.; Lee, C.; Dettlaff-Weglikowska, U. Solvent-Dependent Thermoelectric Properties of PTB7 and Effect of 1,8-Diiodooctane Additive. *Crystals* **2017**, *7*, 292. [CrossRef]

18. Kim, S.S.; Bae, S.; Jo, W.H. Performance enhancement of planar heterojunction perovskite solar cells by n-doping of the electron transporting layer. *Chem. Commun.* **2015**, *51*, 17413–17416. [CrossRef] [PubMed]

19. Wei, P.; Oh, J.H.; Dong, G.; Bao, Z. Use of a 1*H*-Benzoimidazole Derivative as an *n*-Type Dopant and to Enable Air-Stable Solution-Processed *n*-Channel Organic Thin-Film Transistors. *JACS* **2010**, *132*, 8852–8853. [CrossRef] [PubMed]

20. Gao, X.; Uehara, K.; Klug, D.D.; Tse, J.S. Rational design of high-efficiency thermoelectric materials with low band gap conductive polymers. *Comput. Mater. Sci.* **2006**, *36*, 49–53. [CrossRef]

21. Wuesten, J.; Ziegler, C.; Ertl, T. Electron transport in pristine and alkali metal doped perylene-3,4,9,10-tetracarboxylicdianhydride (PTCDA) thin films. *Phys. Rev. B* **2006**, *74*, 125205. [CrossRef]

22. Heremans, J.P.; Jovovic, V.; Toberer, E.S.; Saramat, A.; Kurosaki, K.; Charoenphakdee, A.; Yamanaka, S.; Snyder, G.J. Enhancement of thermoelectric efficiency in PbTe by distortion of the electronic density of states. *Science* **2008**, *25*, 554. [CrossRef] [PubMed]

23. Martin, J.; Wang, L.; Chen, L.D.; Nolas, G.S. Enhanced Seebeck coefficient through energy-barrier scattering in PbTe nanocomposites. *Phys. Rev. B* **2009**, *79*, 115311. [CrossRef]

© 2018 by the authors. Licensee MDPI, Basel, Switzerland. This article is an open access article distributed under the terms and conditions of the Creative Commons Attribution (CC BY) license (http://creativecommons.org/licenses/by/4.0/).

crystals

MDPI

Article

Solvent-Dependent Thermoelectric Properties of PTB7 and Effect of 1,8-Diiodooctane Additive

Mina Rastegaralam [1], Changhee Lee [1],* and Urszula Dettlaff-Weglikowska [2],*

[1]   Department of Electrical and Computer Engineering, Inter-University Semiconductor Research Center, Seoul National University, 1 Gwanak-ro, Gwanak-gu, Seoul 08826, Korea; mn.rstgrlm@gmail.com
[2]   Materials Science Consulting and Management, 22113 Oststeinbek, Moellner Landst. 97B, Germany
*   Correspondence: chlee7@snu.ac.kr (C.L.); udettlaff.w@gmail.com (U.D.-W.)

Academic Editor: George S. Nolas
Received: 19 August 2017; Accepted: 26 September 2017; Published: 29 September 2017

Abstract: Conjugated polymers are considered for application in thermoelectric energy conversion due to their low thermal conductivity, low weight, non-toxicity, and ease of fabrication, which promises low manufacturing costs. Here, an investigation of the thermoelectric properties of poly({4,8-bis[(2-ethylhexyl)oxy]benzo [1,2-b:4,5-b′] dithiophene-2,6-diyl}{3-fluoro-2-[(2-ethylhexyl) carbonyl]thieno[3,4-b] thiophenediyl}), commonly known as PTB7 conjugated polymer, is reported. Samples were prepared from solutions of PTB7 in three different solvents: chlorobenzene, 1,2-dichlorobenzene, and 1,2,4-trichlorobenzene, with and without 1,8-diiodooctane (DIO) additive. In order to characterize their thermoelectric properties, the electrical conductivity and the Seebeck coefficient were measured. We found that, by increasing the boiling point of the solvent, both the electrical conductivity and the Seebeck coefficient of the PTB7 samples were simultaneously improved. We believe that the increase in mobility is responsible for solvent-dependent thermoelectric properties of the PTB7 samples. However, the addition of DIO changes the observed trend. Only the sample prepared from 1,2,4-trichlorobenzene showed a higher electrical conductivity and Seebeck coefficient and, as a consequence, improved power factor in comparison to the samples prepared from chlorobenzene and 1,2-dichlorobenzene.

Keywords: electrical conductivity; Seebeck coefficient; power factor; PTB7 polymer

---

1. Introduction

Thermoelectric materials are capable of the solid-state conversion between thermal and electrical energy. Such materials have attracted much attention due to their great potential for applications in power generation and refrigeration [1,2]. Thermoelectric materials are used in generating power from waste heat and in on-chip and larger-scale cooling modules [3–7]. Thermoelectric generators offer a number of advantages compared to other direct current sources of power. They usually have a compact module structure that does not require any moving parts [8].

The performance of thermoelectric materials is determined by a dimensionless quantity called the thermoelectric figure of merit ZT expressed by: $ZT = S^2\sigma T/\kappa$, where S, σ, T, and κ represent the Seebeck coefficient, electrical conductivity, absolute temperature, and thermal conductivity, respectively.

In order to improve the figure of merit, it is necessary to increase the Seebeck coefficient and electrical conductivity simultaneously. However, it is a challenging task, as both parameters are inversely correlated. The electrical conductivity is calculated from $\sigma = en\mu$, with n being the carrier concentration, μ the carrier mobility, and e the electron charge. At the same time, by increasing n, the electrical conductivity of the material increases, while the Seebeck coefficient decreases. An alternative strategy is the enhancement of mobility while maintaining a constant n to increase electrical conductivity and Seebeck coefficient simultaneously [9].

Inorganic conductors and semiconductors are efficient thermoelectric materials, but they are associated with issues such as high production cost, toxicity, and scarcity [10]. In contrast, organic and polymeric semiconductors are advantageous over inorganic materials due to their relatively easy use, low thermal conductivity, low weight, non-toxicity, and the established thin layer technology, promising low fabrication costs. In addition to conventional polymers, conjugated polymers may also be used as thermoelectric materials since they have advantageous characteristics of conventional polymers. The low thermal conductivity, κ, of conjugated polymers (two or three orders of magnitude lower than in inorganic semiconductors) is a major factor when considering conjugated polymers for thermoelectric applications, besides their ease of fabrication via simple solution processability [8,11–13]. The fact that most conducting polymers can be manufactured in the form of thin films over large areas is a potential benefit for multilayered polymer thermoelectric modules. Thin films require less material than bulk, and can be more easily processed over large surfaces. Both factors contribute to lower fabrication costs. In addition, operating thermoelectric devices are usually modules composed of alternatively layered films of p- and n-type semiconducting materials. Therefore, the solution-processed thin film technology is beneficial for organic thermoelectric materials. If polymer thermoelectrics are produced on a large scale, the cost could be much lower than currently produced bismuth telluride thermoelectrics (~$7/watt) [14], which is the most commonly used material for thermoelectric applications with a ZT value close to 1.

To date, few conducting polymers have been studied for their thermoelectric properties [13,15–18]; therefore, samples of PTB7 polymer were prepared and investigated.

PTB7 is a p-type semiconducting polymer that is used in organic field effect transistors. This conjugated polymer is also widely used in bulk heterojunction polymer solar cells as a donor mixed with an acceptor material in the blend of their active layer. In an effort to improve the efficiency of PTB7-based polymer solar cells, DIO was added to the blend of donor-acceptor materials, giving power conversion efficiency (PCE) of 7.4%. This was the first polymer solar cell to show a PCE over 7% [19]. However, when DIO was mixed with only PTB7, it had, depending on the solvent, different effects on the polymer [20]. As has been shown by Guo et al., only the films made from 1,2,4-trichlorobenzene exhibited an improved crystallinity after addition of DIO compared to chlorobenzene and 1,2-dichlorobenzene [20].

Here, we explored the effect of different solvents and DIO on the thermoelectric properties of PTB7samples fabricated via drop casting. The following sections focus on the preparation of the samples, determination of their electrical conductivity, the Seebeck coefficient, charge carrier mobility, and a discussion of the results.

2. Materials and Methods

PTB7 and DIO were purchased from 1-Material and Tokyo Chemical Industry, respectively, and solvents were purchased from Sigma Aldrich. All materials were used as received. The PTB7 solutions were prepared by dissolving 15 mg of PTB7 in 1 mL (1000 μL) of chlorobenzene (CB), 1,2-dichlorobenzene (DCB), and 1,2,4-trichlorobenzene (TCB) under glove box filled with argon. For solutions containing DIO, 15 mg of PTB7 was dissolved in a mixture of 970 μL of solvent and 30 μL of DIO and stirred for more than 12 hours. Samples were formed by drop casting the solutions on glass substrates, which were cleaned via sonication in acetone, isopropyl alcohol, and water before use. The thickness of each sample was, on average, 10 μm, which was determined by a profilometer. The electrical conductivity was measured via the four-probe method using a Keithley SMU237 source measurement unit. The Hall mobility of the samples was measured in a magnetic field of 0.5 T at room temperature using the Van der Pauw method. The Seebeck coefficient was calculated from $S = -\Delta V / \Delta T$, where ΔV was the thermoelectric voltage generated along the sample when it was subjected to the temperature difference ΔT, as shown schematically in Figure 1. Two K type thermocouples were used to measure the temperature on both ends of the samples. From each solution, four samples were prepared, and electrical conductivity, carrier mobility, and the Seebeck coefficient were measured five times for each sample.

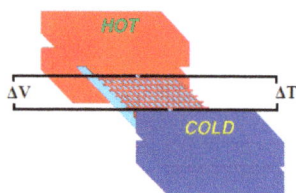

Figure 1. Schematic presentation of the Seebeck coefficient determination.

3. Results and Discussion

Figure 2 shows the molecular structure of the applied chemical materials, and Table 1 presents the boiling points of the solvents and DIO. Among these three solvents, 1,2,4-trichlorobenzene (TCB) has the highest boiling point (214) and chlorobenzene (CB) has the highest vapor pressure and the lowest boiling point (131). As shown in Figure 3, for both types of the samples (with and without DIO), the electrical conductivity value of the samples made of TCB is higher than two other samples made of DCB and CB.

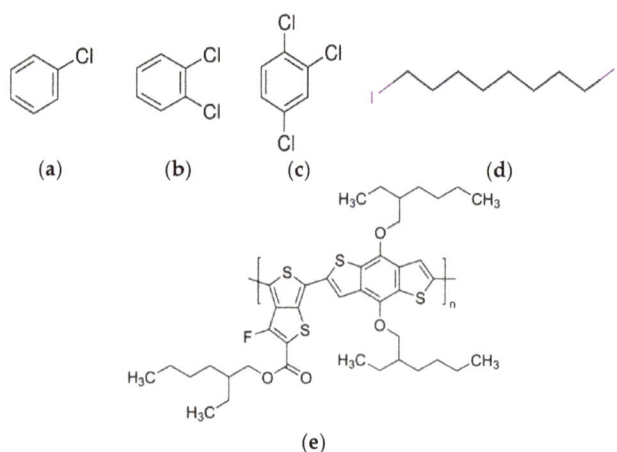

Figure 2. Molecular structure of (**a**) chlorobenzene (CB), (**b**) 1,2-dichlorobenzene (DCB), (**c**) 1,2,4-trichlorobenzene (TCB), (**d**) DIO (**e**) PTB7.

Table 1. Boiling point of the used solvents and 1,8-diiodooctane.

Material	Boiling Point (°C)
1,2,4-Trichlorobenzene ($C_6H_3Cl_3$)	214
1,2-Dichlorobenzene ($C_6H_4Cl_2$)	180
Chlorobenzene (C_6H_5Cl)	131
1,8-Diiodooctane($C_8H_{16}I_2$)	167–169

Considering the boiling point of the solvents, it is clear that samples made of a higher boiling point solvent need a longer time for drying to complete evaporation of the solvent. Slow evaporation of the solvent allows for better ordered molecular packing of the PTB7 polymer and therefore, its ability to transport charge carriers improves, leading to increased electrical conductivity [21].

As shown in Figure 3, the values of conductivities are 0.5 S/m, 0.8 S/m and 1.2 S/m for samples made of CB, DCB, and TCB without DIO, respectively. Furthermore, we observed that the addition of DIO improved the electrical conductivity only in the samples made of TCB, and failed to enhance the electrical conductivity of the samples prepared from CB and DCB. There seems to be a synergic effect in the interaction between the polymer, TCB, and DIO. However, in the solvents that evaporate faster, the introduction of DIO (which remains in the assembled polymer fibers) increases the impurity scattering due to collisions between charge carriers and DIO molecules, which may be the reason for decreasing the carrier mobility and as a consequence, electrical conductivity. By looking at the conductivity values, it is obvious that the addition of DIO has the worst effect on the conductivity of the sample prepared from chlorobenzene. The electrical conductivity values for samples containing DIO are as follows: 0.35 S/m, 0.6 S/m, and 1.78 S/m for samples made of CB, DCB and TCB, respectively.

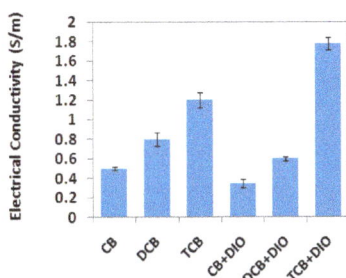

Figure 3. Electrical conductivity values of PTB7 samples prepared from different solvents with and without DIO.

The nature of charge carriers in the PTB7 samples was determined by thermoelectric power measurement. Corresponding Seebeck coefficients of the samples are shown in Figure 4. The positive Seebeck coefficient values indicate that the major charge carriers are holes, confirming PTB7 as a p-type semiconductor. The diagram reflects that the Seebeck coefficients increase with increasing boiling point of the solvent for both types of samples (with and without DIO) as was already observed in the electrical conductivity measurements. For both types of the samples, the highest Seebeck coefficient value was obtained from the samples made of TCB. This simultaneous enhancement in electrical conductivity and the Seebeck coefficient is attributed to an increase in the carrier mobility of the samples, in which a solvent with a higher boiling point was used. Table 2 presents the Hall mobility values of the samples. The changes in the mobility of the samples prepared from different solvents closely correlate with the corresponding measurements of electrical conductivity and the Seebeck coefficient.

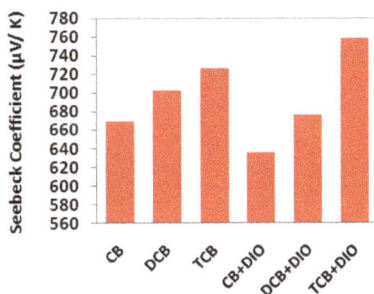

Figure 4. Seebeck coefficient values of PTB7 samples prepared from different solvents with and without DIO.

Table 2. Mobility of the samples.

Sample	Mobility (cm^2/Vs)
PTB7: TCB	1.09×10^{-1}
PTB7: TCB + DIO	1.69×10^{-1}
PTB7: DCB	7.15×10^{-2}
PTB7: DCB + DIO	5×10^{-2}
PTB7: CB	3.9×10^{-2}
PTB7: CB + DIO	2.66×10^{-2}

Upon addition of DIO, Seebeck coefficient values in the samples made of CB and DCB decrease, showing that DIO plays a negative role in increasing mobility in CB and DCB samples. The Seebeck coefficient values of the samples are: 670 µV/K, 702 µV/K and 726 µV/K for CB, DCB and TCB samples without DIO and 636 µV/K, 676 µV/K and 758 µV/K with DIO, respectively.

Figure 5 shows the AFM images of the PTB7 samples. As seen in Figure 5a–c, all the neat PTB7 samples are rather homogeneous as is expected for homopolymer films. For both types of samples, surface roughness decreases when the boiling point of the solvent increases, which is consistent with increasing the mobility. As shown in Figure 5d–f, polymer aggregation is clearly visible when DIO is added. Obviously, the addition of DIO decreases the solubility of polymer in the neat solvents leading to aggregation of polymer fibers and formation of lateral structures in the ternary system (polymer, solvent, DIO). Considering the volatilities of the used solvents that decrease from CB (131 °C) to DB (180 °C) and TB (214 °C), the observed differences in the structure can be explained. The root-mean-square (RMS) surface roughness was found to be 1.1, 0.92 and 0.82 for CB, DCB and TCB samples without DIO and 3.8, 3.2 and 0.7 for CB, DCB and TCB samples with DIO, respectively.

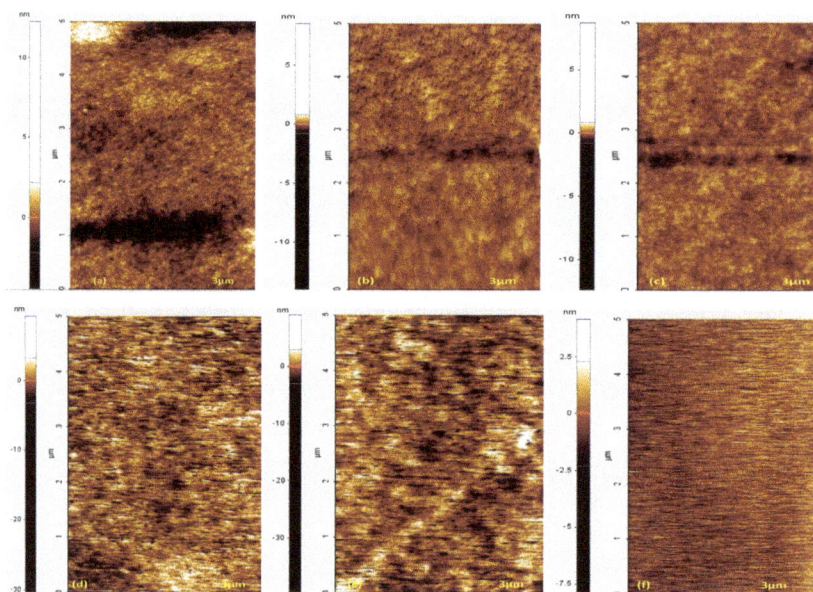

Figure 5. AFM images of PTB7 samples prepared from different solvents with and without DIO: (**a**) CB; (**b**) DCB; (**c**) TCB; (**d**) CB-DIO; (**e**) DCB-DIO; (**f**) TCB-DIO.

Figure 6 shows the thermoelectric power factor values of our samples calculated from P = S^{2}σ, where S is the Seebeck coefficient value and σ is the electrical conductivity. For both types of samples (with and without DIO), the highest power factor was obtained from the 1,2,4-trichlorobenzene sample that shows among these 3 solvents, 1,2,4-trichlorobenzene is the best one for preparing PTB7 solutions in order to obtain samples with higher thermoelectric power factor and addition of DIO has a positive effect on that, while reduces power factor of the CB and DCB samples. The calculated values of the power factor are: 0.00224 (μWK^{-2}cm^{-1}), 0.0039 (μWK^{-2}cm^{-1}) and 0.0063 (μWK^{-2}cm^{-1}) for samples made of CB, DCB and TCB without DIO and 0.00141 (μWK^{-2}cm^{-1}), 0.00274 (μWK^{-2}cm^{-1}) and 0.01 (μWK^{-2}cm^{-1}) with DIO, respectively.

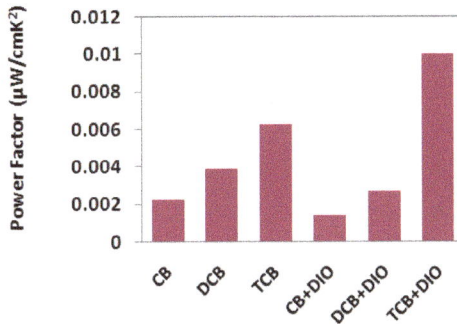

Figure 6. Power factor values of PTB7 samples prepared from different solvents with and without DIO.

It is worth noting that both the electrical conductivity and the Seebeck coefficient and therefore, the power factor values of PTB7 samples are higher than some previously reported conducting polymers such as poly [3-hexylthiophene] (P3HT) [13], poly [N-90-heptadecanyl-2,7-carbazole-alt-5,5-(40,70-di-2-thienyl-20,10,3-benzothiadizole)] (PCDTBT) [16], poly carbazolenevinylene derivative, poly paraphenylene, poly *p*-phenylenevinylene [22] and [6,6]-phenyl-C61-butyric acid methyl ester (PCBM) [23]. However, we believe that any changes in chemical structure of the conducting polymers, lead to a modification in their electrical conductivity and influence the thermoelectric effect. The replacement of the functional groups by different side chains in PTB7 will affect the assembly of polymer fibers in the solution-processed layers and polymer solubility and thus, modify their thermoelectric behavior. For this purpose, detailed computer simulations followed by experimental studies are required to find out which kind of side chains will have a positive effect on the thermoelectric properties of the modified PTB7.

Assuming that the thermal conductivity of the PTB7 samples is comparable to the value provided for polypyrrole (0.1) [22], the figure of merit (ZT) values for our samples are estimated to be 0.000673, 0.00118 and 0.0019 for samples made of CB, DCB and TCB without and 0.000424, 0.00082 and 0.003 with DIO, respectively, which is still some orders of magnitude lower than that of bismuth telluride. However, thin layer technology established for conductive polymers enables the fabrication of multi-layered structures. In future work, we aim to investigate multiple element modules composed of alternatively layered films with PTB7 as an electron-donating polymer and films of an n-type semiconducting polymer. The thermoelectric voltage generated by such modules would be the sum of contributions from each layer, resulting in increased power output.

4. Conclusions

We investigated the effect of different solvents and 1,8-diiodooctane additive on the thermoelectric properties of PTB7 polymer. Our data demonstrate that a suitable solvent leads to the improvement of thermoelectric properties and the enhancement of the figure of merit. Dissolving PTB7 in a solvent with a higher boiling point leads to obtain higher electrical conductivity and Seebeck coefficient values due to a better ordering of polymer fibers which is beneficial for charge carrier mobility. We observed that addition of DIO improves the electrical conductivity and the Seebeck coefficient of PTB7 only in the sample prepared from 1,2,4-trichlorobenzene, while for the samples prepared from chlorobenzene and 1,2-dichlorobenzene, DIO has a negative effect on the thermoelectric properties reflected by the power factor.

Acknowledgments: This work was supported by the National Research Foundation of Korea (NRF-2016R1A2B3009301) and LG Chem, Ltd.

Author Contributions: Mina Rastegaralam conceived, designed and performed the experiments; Urszula Dettlaff-Weglikowska coordinated scientific work and interpreted the experimental results and Changhee Lee wrote the paper. All authors read and approved the final version of the manuscript to be submitted.

Conflicts of Interest: The authors declare no conflict of interest.

References

1. Poudel, B.; Hao, Q.; Ma, Y.; Lan, Y.; Minnich, A.; Yu, B.; Yan, X.; Wang, D.; Muto, A.; Vashaee, D.; et al. High-Thermoelectric Performance of Nanostructured Bismuth Antimony Telluride Bulk Alloys. *Science* **2008**, *320*, 634–638. [CrossRef] [PubMed]
2. Zhao, W.; Fan, S.; Xiao, N.; Liu, D.; Tay, Y.Y.; Yu, C.; Sim, D.; Hng, H.H.; Zhang, Q.; Boey, F.; et al. Flexible Carbon Nanotube Papers with Improved Thermoelectric Properties. *Energy Environ. Sci.* **2012**, *5*, 5364–5369. [CrossRef]
3. Adroja, N.; Mehta, S.B.; Shah, P. Review of thermoelectricity to improve energy quality. *Int. J. Emerg. Technol. Innov. Res.* **2015**, *2*, 847–850.
4. Sharp, J.; Bierschenk, J.; Lyon, H.B. Overview of Solid-State Thermoelectric Refrigerators and Possible Applications to On-Chip Thermal Management. *Proc. IEEE* **2006**, *94*, 1602–1612. [CrossRef]
5. Bell, L.E. Cooling, heating, generating power, and recovering waste heat with thermoelectric systems. *Science* **2008**, *321*, 1457–1461. [CrossRef] [PubMed]
6. Snyder, G.J.; Toberer, E.S. Complex thermoelectric materials. *Nat. Mater.* **2008**, *7*, 105–114. [PubMed]
7. Karni, J. Solar energy: The thermoelectric alternative. *Nat. Mater.* **2011**, *10*, 481–482. [CrossRef] [PubMed]
8. Xu, L.; Liu, Y.; Chen, B.; Zhao, C.; Lu, K. Enhancement in thermoelectric properties using a P-type and N-type thin-film device structure. *Polym. Compos.* **2013**, *34*, 1728–1734. [CrossRef]
9. Petsagkourakis, I.; Pavlopoulou, E.; Portale, G.; Kuropatwa, B.A.; Dilhaire, S.; Fleury, G.; Hadziioannou, G. Structurally-driven enhancement of thermoelectric properties within poly 3,4-ethylenedioxythiophene) thin films. *Sci. Rep.* **2016**, *6*, 30501. [CrossRef] [PubMed]
10. Dubey, N.; Leclerc, M. Conducting Polymers: Efficient Thermoelectric Materials. *J. Polym. Sci. B Polym. Phys.* **2011**, *49*, 467–475.
11. Lodha, A.; Singh, R. Prospects of manufacturing organic semiconductor- based integrated circuits. *IEEE Trans. Semicond. Manuf.* **2001**, *14*, 281–296. [CrossRef]
12. Goodson, K.E.; Ju, Y.S. Heat conduction in novel electronic films. *Annu. Rev. Mater. Sci.* **1999**, *29*, 261–293. [CrossRef]
13. Sun, J.; Yeh, M.L.; Jung, B.J.; Zhang, B.; Feser, J.; Majumdar, A.; Katz, H.E. Simultaneous increase in Seebeck coefficient and conductivity in a doped poly (alkylthiophene) blend with defined density of states. *Macromolecules* **2010**, *43*, 2897–2903. [CrossRef]
14. Hewitt, C.A.; Kaiser, A.B.; Roth, S.; Craps, M.; Czerw, R.; Carroll, D.L. Multilayered carbon nanotube/polymer composite based thermoelectric fabrics. *Nano Lett.* **2012**, *12*, 1307–1310. [CrossRef] [PubMed]
15. Ma, W.; Shi, K.; Wu, Y.; Lu, Z.Y.; Liu, H.Y.; Wang, J.Y.; Pei, J. Enhanced molecular packing of a conjugated polymer with high organic thermoelectric power factor. *ACS Appl. Mater. Interfaces* **2016**, *8*, 24737–24743. [CrossRef] [PubMed]
16. Maiz, J.; Rojo, M.M.; Abad, B.; Wilson, A.A.; Nogales, A.; Borca-Tasciuc, D.A.; Borca-Tasciuc, T.; Martín-González, M. Enhancement of thermoelectric efficiency of doped PCDTBT polymer films. *RSC Adv.* **2015**, *5*, 66687–66694. [CrossRef]
17. Mateeva, N.; Niculescu, H.; Schlenoff, J.; Testardi, L.R. Correlation of Seebeck coefficient and electric conductivity in polyaniline and polypyrrole. *J. Appl. Phys.* **1998**, *83*, 3111–3117. [CrossRef]

18. Toshima, N. Conductive polymers as a new type of thermoelectric material. *Macromol. Symp.* **2002**, *186*, 81–86. [CrossRef]

19. Liang, Y.; Xu, Z.; Xia, J.; Tsai, S.; Wu, Y.; Li, G.; Ray, C.; Yu, L. For the bright future-bulk heterojunction polymer solar cells with power conversion efficiency of 7.4%. *Adv. Mater.* **2010**, *22*, E135–E138. [CrossRef] [PubMed]

20. Guo, S.; Herzig, E.M.; Naumann, A.; Tainter, G.; Perlich, J.; Muller-Buschbaum, P. Influence of solvent and solvent additive on the morphology of PTB7 films probed via X-ray scattering. *J. Phys. Chem. B* **2014**, *118*, 344–350. [CrossRef] [PubMed]

21. Foote, A.L. Investigation of Solvent-Dependent Properties of Donor and Acceptor Materials for Photovoltaic Applications. Available online: http://corescholar.libraries.wright.edu/etd_all/1451/ (accessed on 29 July 2015).

22. Xuan, Y.; Liu, X.; Desbief, S.; Leclère, P.; Fahlman, M.; Lazzaroni, R.; Berggren, M.; Cornil, J.; Emin, D.; Crispin, X. Thermoelectric properties of conducting polymers: The case of poly(3-hexylthiophene). *Phys. Rev. B* **2010**, *82*, 115454. [CrossRef]

23. Gao, F.; Liu, Y.; Xiong, Y.; Wu, P.; Hu, B.; Xu, L. Fabricate organic thermoelectric modules use modified PCBM and PEDOT:PSS materials. *Front. Optoelectron.* **2017**, *10*, 117–123. [CrossRef]

© 2017 by the authors. Licensee MDPI, Basel, Switzerland. This article is an open access article distributed under the terms and conditions of the Creative Commons Attribution (CC BY) license (http://creativecommons.org/licenses/by/4.0/).

crystals

MDPI

Article

Microstructure Evolution of Ag-Alloyed PbTe-Based Compounds and Implications for Thermoelectric Performance

Tom Grossfeld [1], Ariel Sheskin [2], Yaniv Gelbstein [3] and Yaron Amouyal [2,*]

[1] The Nancy and Stephen Grand Technion Energy Program (GTEP), Technion-Israel Institute of Technology, Haifa 32000, Israel; tomgrossfeld@gmail.com
[2] Department of Materials Science and Engineering, Technion-Israel Institute of Technology, Haifa 32000, Israel; ariel.s@campus.technion.ac.il
[3] Department of Materials Engineering, Faculty of Engineering Sciences, Ben-Gurion University of the Negev, Beer Sheva 84105, Israel; yanivge@bgu.ac.il
* Correspondence: amouyal@technion.ac.il

Academic Editor: Shujun Zhang
Received: 17 August 2017; Accepted: 12 September 2017; Published: 18 September 2017

Abstract: We investigate the microstructure evolution of Ag-alloyed PbTe compounds for thermoelectric (TE) applications with or without additions of 0.04 at. % Bi. We control the nucleation and temporal evolution of Ag_2Te-precipitates in the PbTe-matrix applying designated aging heat treatments, aiming to achieve homogeneous dispersion of precipitates with high number density values, hypothesizing that they act as phonon scattering centers, thereby reducing lattice thermal conductivity. We measure the temperature dependence of the Seebeck coefficient and electrical and thermal conductivities, and correlate them with the microstructure. It is found that lattice thermal conductivity of PbTe-based compounds is reduced by controlled nucleation of Ag_2Te-precipitates, exhibiting a number density value as high as 2.7×10^{20} m^{-3} upon 6 h aging at 380 °C. This yields a TE figure of merit value of ca. 1.4 at 450 °C, which is one on the largest values reported for n-type PbTe compounds. Subsequent aging leads to precipitate coarsening and deterioration of TE performance. Interestingly, we find that Bi-alloying improves the alloys' thermal stability by suppressing microstructure evolution, besides the role of Bi-atoms as electron donors, thereby maintaining high TE performance that is stable at elevated service temperatures. The latter has prime technological significance for TE energy conversion.

Keywords: thermoelectric materials; PbTe; thermal conductivity; phonon scattering; phase transformations; microstructure evolution

1. Introduction

Lead-telluride (PbTe) based compounds are common thermoelectric (TE) materials for the mid-temperature range (600–800 K) [1]. These narrow band-gap semiconductors (ca. 0.3 eV at room temperature), which have been thoroughly investigated, offer unique combination of high Seebeck coefficient, S, with relatively high electrical conductivity, σ, and low thermal conductivity, κ. Owing to this combination, single-phase PbTe exhibits a maximum dimensionless TE *figure of merit* (ZT) value of ca. 0.8 [2,3], which can approach ca. 2.0 owing to doping and nanostructuring [4–6]. The lattice thermal conductivity of PbTe is typically 2.2 $Wm^{-1}K^{-1}$ at room temperature [7,8], which is yet far away from the theoretical alloy limit [9]. This opens up prospects for tuning the TE performance by introducing lattice defects.

Reports indicate that phonons having low- to mid-range frequencies (i.e., mid- to long-wavelength range) can be significantly scattered by nanostructured features, such as precipitates [10], whereas high-frequency phonons (i.e., of short wavelengths) are scattered by point defects mainly, e.g., solute atoms and vacancies [11]. In this sense, besides achieving direct enhancement of the TE performance, understanding basic physical behavior of lattice thermal conductivity of two-phase alloys based on PbTe is a grand challenge by itself. This may be achieved by distinction between the effects of nano-particles volume fraction, number density, and average size—on the one hand, and those of the matrix composition—on the other hand [12,13]. Thorough investigations of the microstructure formed in PbTe-based ternary and quaternary systems, including its temporal evolution and effects on TE behavior, have been reported [14–21]. Such effects were particularly studied for Ag-alloyed PbTe compounds as well [22–26]. Nevertheless, neither of these studies introduces systematic investigations of the effects of second-phase precipitates' volume fraction, number density, and average size, as well as their temporal evolution, on TE transport properties of PbTe-based alloys.

In this study, we focus on Ag-alloyed PbTe-based systems having the potential to form Ag_2Te-precipitates dispersed in the PbTe-based solid solution. We perform *controlled* aging heat treatments at different temperatures and durations to enhance nucleation of Ag_2Te-precipitates for directly reducing lattice thermal conductivity. It is hypothesized that aging should yield increased nucleation of Ag_2Te-precipitates up to a time limit where over-aging processes take place [27], which is accompanied by changes in the PbTe-based matrix average composition. The side-effects on other TE transport properties and, in general, on TE performance are also examined. This approach was demonstrated by us recently for a ZnO-based system, with clear conclusions regarding TE performance [28]. In this context, we cope with two main challenges that concern the materials synthesis process, which are optimization of heat treatment conditions and selection of the processing routes, such as powder pressing and fast cooling, and highlight the essence of these factors in determination of the TE performance. Herein, knowledge of the temporal evolution of PbTe-matrix/Ag_2Te-precipitates system's microstructure upon aging will serve for improvement of the TE performance of PbTe-based compounds.

2. Experimental Procedure

2.1. Materials Synthesis

In this study we report on two classes of materials; the first one is synthesized by casting, and the second one by hot-pressing.

As-cast (AC) compounds are synthesized from pure elemental Pb powder (99.96%, Riedel-de Haën®, Hanover, Germany), Te ingots (99.99%, STREM CHEMICALS®, Newburyport, MA, USA), and Ag shots (99.999%, Alfa Aesar®) by mixing in the appropriate molar ratios to obtain the average composition of $(PbTe)_{0.95}(Ag_2Te)_{0.05}$. We choose this composition since it is within the single-phase regime at temperatures above 600 °C, and is expected to decompose into a two-phase mixture, namely Ag_2Te + PbTe, in a solid-state precipitation process at lower temperatures. The pure Pb-, Te-, and Ag-raw materials are poured into a 12.5 dia. quartz ampoule, which is evacuated and refilled with a 120 torr Ar-7% H_2 gas mixture to avoid oxidation. The sealed ampoule is subsequently heated to 1000 °C for 2 h in a vertical programmable tube furnace to enable melting. Termination of the melting is performed by quenching in iced-water bath, followed by annealing at 700 °C for 24 h to homogenize the solid solution at the single-PbTe-phase regime and, finally, quenching in an iced-water bath. In order to precipitate the Ag_2Te-phase from the solid solution, the ingot is sliced to disk-shape specimens, ca. 2 mm thick, and aged at 400 and 450 °C for different durations, as listed in Table 1, in sealed evacuated ampoules to enable us obtain Ag_2Te-precipitates of different sizes and number densities dispersed in the PbTe-based solid solution.

Hot-pressed (HP) specimens of the $(PbTe)_{0.97}(Ag_2Te)_{0.03}$ base composition are prepared in two batches; the first one does not contain Bi, and the second one contains 0.04 at. % Bi. These compositions and synthesis method are chosen in a way that allows us to prepare single-phase supersaturated specimens that are mechanically robust to sustain subsequent TE measurements. First, granules of

Pb, Te, Ag, and Bi (99.5%, LOBA CHEMIE®, Mumbai, India) are mixed in the appropriate fractions and undergo repeated arc melting processes. Then, the solid solution is grinded into fine powder and heated up to 650 °C (within the single-phase regime) under a pressure of 21 MPa for 0.5 h in a 30 mm dia. die, and then cut into 12.5 mm dia. disks. Then, the specimens are aged at 380 °C for different durations, as listed in Table 1, in sealed and evacuated ampoules under a 120 torr Ar-7% H_2 atmosphere to initiate nucleation of the Ag_2Te-phase, followed by iced-water quenching.

Table 1. A list of the materials' compositions, synthesis conditions, and heat treatments carried out for each specimen.

Material	Aging Temperature [°C]	Aging Times [h]
As-cast (AC) $(PbTe)_{0.95}(Ag_2Te)_{0.05}$	400	0, 2, 8, 32, 72, 106
	450	0, 1, 2, 4, 8, 16, 32
Hot-pressed (HP) $(PbTe)_{0.97}(Ag_2Te)_{0.03}$	380	0, 0.5, 1, 2, 4, 6, 8, 16, 24, 48
Hot-pressed (HP) 0.04 at. % Bi + $(PbTe)_{0.97}(Ag_2Te)_{0.03}$	380	0, 2, 6, 48

2.2. Characterization Methods

2.2.1. Microstructure Analysis

We apply powder X-ray diffraction (XRD) to determine the phases present in the heat treated ingots. Measurements are carried out using a SmartLab® XRD diffractometer (Rigaku, The Woodlands, TX, USA) with an angular resolution of 0.02°, applying Cu-K_α radiation at the angular range of $2\theta = 20–120°$. Data are collected with angular scanning resolution of 0.03° per step. Microstructure characterization of the samples is carried out using a Ultra Plus® high-resolution scanning electron microscope (Zeiss, Oberkochen, Germany) (HRSEM) equipped with an 80 mm^2 active area Oxford® SDD electron dispersive spectroscopy (EDS) detector with an energy resolution of 127 eV, equipped with a Schottky field-emission electron gun. In some cases, metallographic polishing of the specimens' surface is carried out by ion milling employing a Helios NanoLab DualBeam® (FEI, Hillsboro, OR, USA) G3 UC dual beam focused ion beam (FIB) to reveal the microstructure features. All micrographs presented in this study are acquired using either secondary or backscattered signals, to be specified further below, operated in the range between 3 and 15 kV. Finally, the precipitate number density, N_v, defined as number per unit volume, is determined based on the two-dimensional HRSEM micrographs showing the Ag_2Te-precipitates in a contrast that differs from that of the PbTe-based matrix, following the methodology introduced by us elsewhere [28].

2.2.2. Thermoelectric Property Measurements

We employ the laser flash analysis (LFA) technique, using an LFA-457 MicroFlash® apparatus (Netzsch, Selb, Germany), to directly measure the thermal diffusivity, α, of disk-shaped specimens at the temperature range of 25 through 600 or 700 °C, depending on the material. The thermal conductivity is then calculated from the relationship:

$$\kappa(T) = \alpha(T) \cdot \rho \cdot C_p(T) \tag{1}$$

where ρ is the material's bulk density as evaluated from the sample dimensions and mass, and its temperature dependence is neglected. C_p is the heat capacity, measured indirectly using the LFA with respect to a standard made of *pyroceram 9606* or pure alumina. Electrical conductivity and Seebeck coefficient values are measured employing an SBA-458 Nemesis® system (Netzsch, Selb, Germany) for specimens having the same geometry as for the LFA. All of the measurements are carried out under flowing Ar as a protective atmosphere.

3. Results

3.1. Microstructure Characterization

We perform XRD analysis for specimens that were hot-pressed prior to aging heat treatments to assure that the material is a PbTe-based solid solution single phase, or does not contain considerable amount of Ag_2Te-precipitates. Figure 1 presents an XRD pattern collected from a hot-pressed specimen of the $(PbTe)_{0.97}(Ag_2Te)_{0.03}$ composition, which did not undergo further aging heat treatments. The fully-indexed XRD pattern implies that the material is a single PbTe-phase.

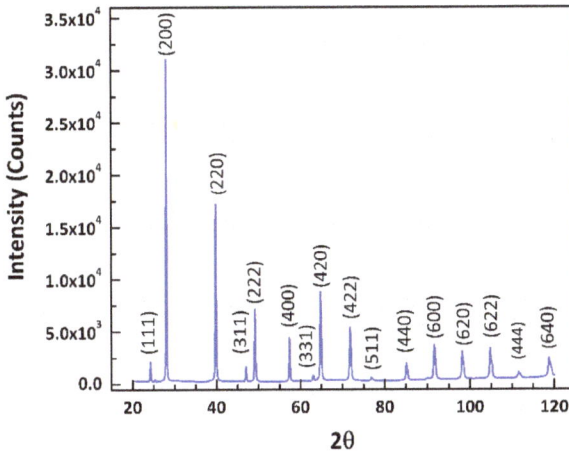

Figure 1. A fully-indexed powder x-ray diffraction (XRD) pattern acquired from the hot-pressed (HP) as-pressed, $(PbTe)_{0.97}(Ag_2Te)_{0.03}$ series, corresponding with the rock-salt PbTe crystal structure (JCPDS number 04-002-0317).

HRSEM analysis is performed for all of the specimens to characterize their microstructures. Figure 2 displays selected HRSEM micrographs taken from the HP series of specimens with or without Bi. It is noteworthy that, first, the presence of elongated Ag_2Te-precipitates is apparent for the specimens that were not aged, although they were iced-water quenched and are expected to comprise a single PbTe solid solution phase. EDS analysis was carried out for additional specimens to validate the Ag_2Te-stoichiometry. The average concentrations inferred from analysis of several large precipitates is 61.7 ± 1.1 at. % Ag; 38.0 ± 1.1 at. % Te; Pb bal. The number densities of these Ag_2Te-precipitates are, however, relatively small. Second, the Bi-free specimens exhibit precipitate number density that increases up to 6 h aging, then decreases up to 48 h aging, whereas the Bi-alloyed specimens exhibit relatively low precipitate number densities that do not change significantly with aging time. For example, quantitative analysis indicates that N_V exceeds values as great as 2.8×10^{17} and 2.7×10^{20} m^{-3} after 6 h aging of the Bi-alloyed and Bi-free materials, respectively. The respective number densities reached after 48 h aging are 8.1×10^{16} and 1.1×10^{19} m^{-3}. It is also shown that the Bi-alloyed specimens preserve the morphology of elongated precipitates over aging time, whereas the precipitate morphology in the Bi-free specimens evolves toward a spheroidal one. This transition takes place between 4 and 6 h aging.

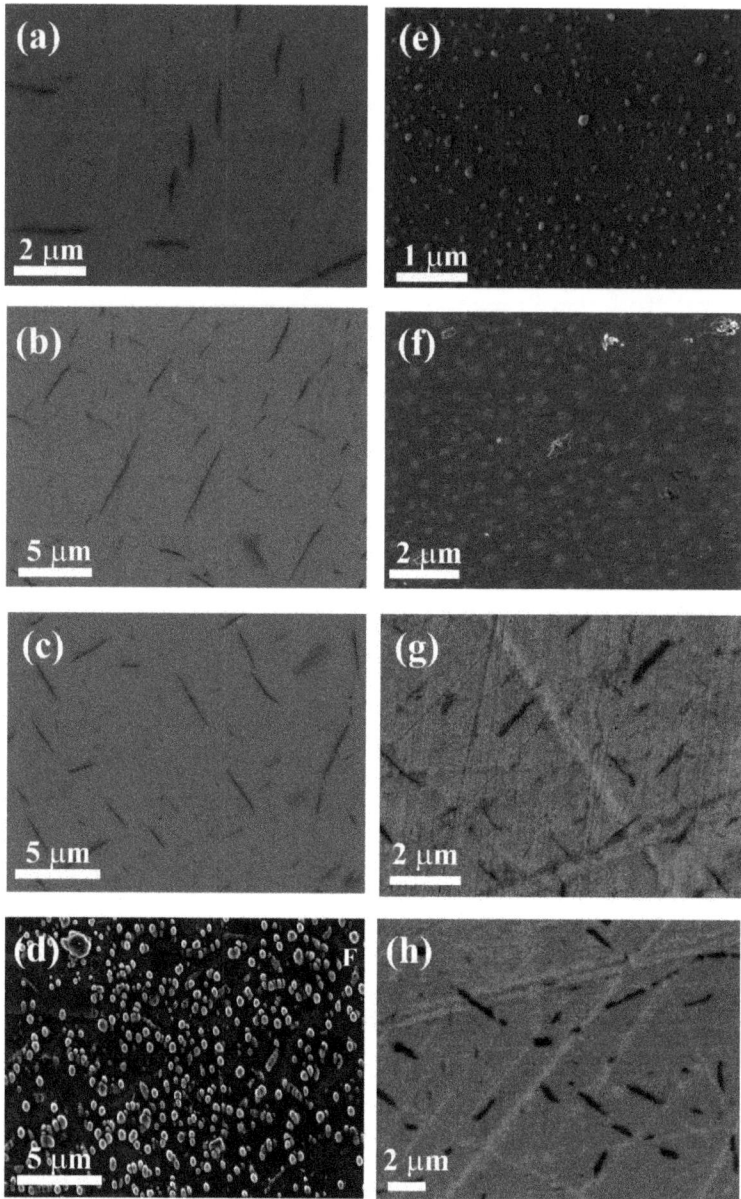

Figure 2. High-resolution scanning electron microscopy (HRSEM) micrographs taken from the polished surfaces of the series of hot-pressed (HP) samples aged at 380 °C for different durations, showing either needle-shaped or spheroidal Ag_2Te-precipitates embedded in a PbTe-based matrix. (**a**) through (**f**) display Bi-free specimens aged for 0, 1, 4, 6, 16, and 48 h, respectively. (**g,h**) display Bi-alloyed specimens aged for 6 and 48 h, respectively. (**a–c,g,h**) are collected applying the back-scattered electrons (BSE) signal, whereas (**d–f**) are collected applying the secondary electrons (SE) signal.

Figure 2 indicates that the Bi-free specimens exhibit strong dependence of the microstructure on aging time, implying that their TE transport coefficients can be modified and controlled more easily compared to the Bi-alloyed ones. To draw conclusions on the microstructure temporal evolution of these specimens on a quantitative basis, we evaluate the number densities of Ag_2Te-precipitates for the Bi-free HP specimens aged at 380 °C, and compare them with those acquired from the AC specimens aged at 400 and 450 °C. The N_V-values for these specimens are plotted against aging time, and are shown in Figure 3.

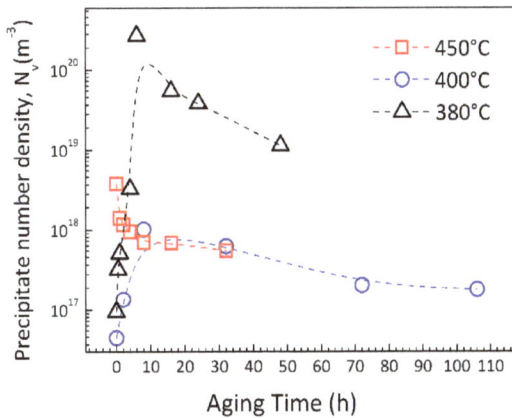

Figure 3. Precipitate number density (N_V) values measured for the different aging times and temperatures, plotted for the $(PbTe)_{0.95}(Ag_2Te)_{0.05}$ as-cast (AC) samples aged at 400 °C (blue circles) and 450 °C (red squares), as well as for the hot-pressed (HP) $(PbTe)_{0.97}(Ag_2Te)_{0.03}$ samples aged at 380 °C (black triangles). For the series of samples aged at 450 °C N_V decreases with aging time, whereas for the samples aged at 400 °C and 380 °C sharp increase of N_V values is indicated for short durations, followed by a decrease.

It is shown that the N_V-values of the AC specimens that were aged at 450 °C decrease with aging time, and reach at an asymptotic value, a behavior typical for coarsening or "over-aging" [27]. The 400 °C- and 380 °C-aged ones, however, exhibit a different trend, which is more desirable for design of TE properties: a maximum value of N_V is achieved for each series.

3.2. Thermoelectric Transport Properties

The temperature dependent electrical conductivity and Seebeck coefficient values are measured for the HP series of specimens aged for different durations, with or without Bi-additions. Selected results are displayed in Figure 4a,b, respectively. The electrical conductivity values exhibit a general trend of increase with increasing temperature, typical for a semiconducting behavior, with an exception for the as-quenched Bi-alloyed material that indicates decrease in the electrical conductivity from a value of ca. 750 $S \cdot cm^{-1}$, typical for a metallic behavior. Also, this general trend is less pronounced for the Bi-free materials aged for 48 h. These trends will be discussed further below.

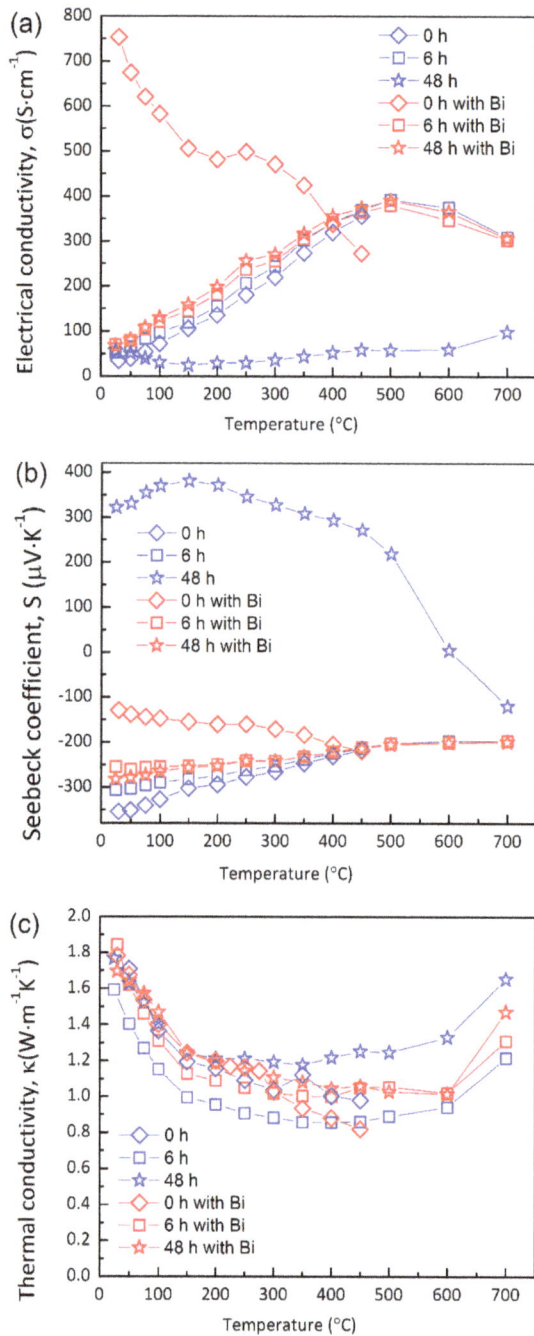

Figure 4. The temperature dependent (**a**) electrical conductivity, (**b**) Seebeck coefficient, and (**c**) thermal conductivity values measured for the hot-pressed (HP) series of specimens aged at 380 °C for 0 (diamonds), 6 (squares), and 48 h (stars) for the Bi-free (blue symbols) and Bi-alloyed (red symbols) materials.

It is also indicated that most of the Bi-doped samples exhibit electrical conductivity values that are slightly higher than those of their undoped counterparts up to 450 °C.

Figure 4b presents the Seebeck coefficient values vs. temperature, in which the trend is opposite to that of the electrical conductivity, as expected. Most samples show decrease in the absolute values of the Seebeck coefficient as the temperature increases up to 450 °C. Additionally, it is shown that all samples except one exhibit negative values, indicating their n-type polarity. As expected, the Bi-doped compounds possess lower absolute values of Seebeck coefficients compared to those of their undoped counterparts, due to increased charge carrier concentrations. Interestingly, at temperatures higher than ca. 450 °C, the curves describing the doped and undoped states converge.

Thermal conductivity values are evaluated based on Equation (1) by direct measurements of thermal diffusivity and room temperature values of the materials' densities, which are listed in Table 2 for the HP series. Selected results appear in Figure 4c. All $\kappa(T)$-curves exhibit similar values and trends.

Table 2. A list of hot-pressed (HP) samples' bulk densities, as measured after the different aging heat treatments.

Aging Time (h)	Bulk Density (g·cm^{-3})	
	Bi-Free	**Bi-Alloyed**
0	8.00	7.86
1	7.65	7.42
2	7.82	7.45
6	7.97	7.37
18	7.85	7.38
48	7.97	7.58

The bulk density is 7.60 g·cm^{-3} on average for the AC series. To evaluate the effects of Ag$_2$Te-precipitates on thermal conductivity via phonon scattering mechanism, it is required to derive the lattice component of thermal conductivity, κ_l. This is implemented applying the following relationship:

$$\kappa = \kappa_e + \kappa_l \tag{2}$$

where κ_e is the electronic component of thermal conductivity, and is evaluated relying on the measured electrical conductivity, Figure 4a, utilizing the Wiedemann-Franz relationship [29]:

$$\kappa_e = L\sigma T \tag{3}$$

where L is evaluated using a semi-empirical expression derived by Kim et al. [30], and is a function of S. It is valid for deviations from the degenerate limit, and was validated for PbTe-based compounds, as well. We calculate L for all samples and temperatures utilizing the measured S-values appearing in Figure 4b.

Figure 5 shows the temperature dependent lattice component thermal conductivity for the (a) Bi-free and (b) Bi-doped HP samples aged for 0, 6, and 48 h. Both types of samples show a similar trend of decreasing thermal conductivity with increasing temperature, which is associated to Umklapp phonon scattering [8]. For temperatures larger than 500 °C the κ_l-values seem to increase with temperature, which deviates from the Umklapp-behavior; this is probably due to phenomena that are not considered by the expression of L, such as bipolar effects [30].

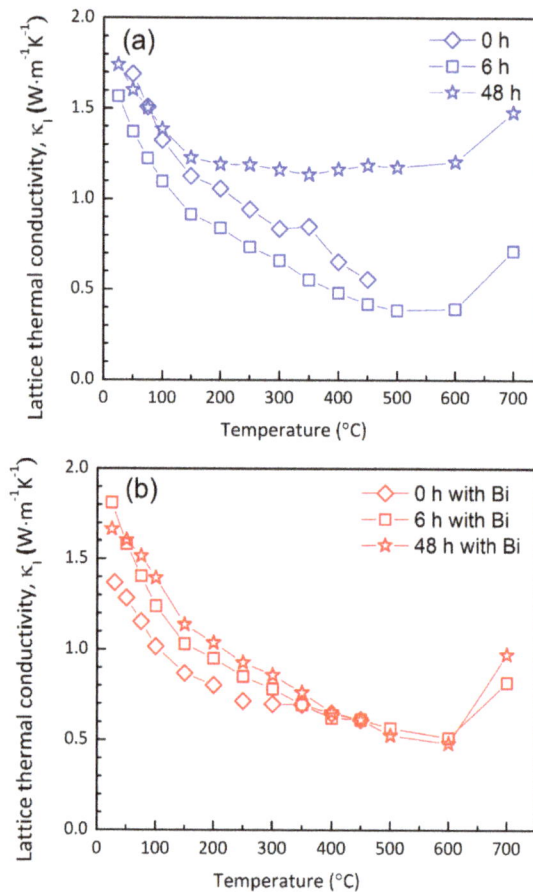

Figure 5. The temperature dependent lattice thermal conductivity values evaluated from the measured thermal and electrical conductivities applying the Wiedemann-Franz relationship for the hot-pressed (HP) series of specimens aged at 380 °C for 0 (diamonds), 6 (squares), and 48 h (stars), for the (**a**) Bi-free (blue symbols) and (**b**) Bi-alloyed (red symbols) materials.

Interestingly, it is indicated that the Bi-free compounds exhibit the lowest κ_l values for 6 h aging, whereas 48 h aging leads to maximum κ_l values. Conversely, the Bi-alloyed compounds exhibit κ_l values with increasing order of their aging times. These trends will be discussed further below.

4. Discussion

4.1. Materials Processing and Microstructure Evolution

It is indicated in Figures 2 and 3 that the materials microstructure, particularly the Ag_2Te-precipitate number density, can be controlled by selecting the appropriate synthesis procedures and aging heat treatments. The specimens aged at 450 °C show an undesirable behavior, in which N_v monotonously decreases with aging time, starting from a relatively large value of ca. 3×10^{19} m^{-3}. This means that the onset of Ag_2Te-precipitates nucleation takes place already during the quenching process. This can be explained by combination of the following factors. First, the PbTe matrix is highly supersaturated with Ag atoms, so that quenching from the single-phase region to two-phase region (below 450 °C) is inadequately fast. Second, the diffusion rate of Ag in the PbTe-matrix might be fast enough to enable the onset of Ag_2Te nucleation just below the PbTe-solvus. Based on this experimental work, as compared to calculations of the expected cooling rates for cylindrical telluride ingots (in direct contact with the quartz wall) conducted by Ikeda et al. [15], we estimate the cooling time of the center of the ingot to be between 30 and 60 s until it reaches room temperature. This is sufficient to enable nucleation of the Ag_2Te precipitates at intermediate temperatures. Ag is known to be a fast diffusing element in PbSe with $D_0 = 7.4 \cdot 10^{-8}$ $m^2 \cdot s^{-1}$ and $Q = 33.77$ kJ·mole^{-1}; valid for the temperature range of 400 through 850 °C [25], where D_0 and Q are the pre-exponential diffusion coefficient and activation energy, respectively. This corresponds to a diffusion length, $\sqrt{Dt}$, of ca. 265 μm for 1 min annealing at 650 °C. Ag diffusion is expected to be faster in PbTe than in PbSe, since the former has a larger lattice constant [31], given that both have the same crystal structure and Ag diffusion should occur in interstitial mechanism [25]. Thus, due to the high diffusivity of Ag, it is reasonable that nanoscale precipitates form during quenching [26]. Lowering the aging temperature from 450 °C to 400 °C enables achieving slower kinetics and lower diffusion rate of Ag atoms in the PbTe matrix at the same time with increased nucleation rate, I, of the Ag_2Te-phase, Equation (4) [27]:

$$I \propto \exp\left(-\frac{\Delta G^* + \Delta G_D}{k_B T}\right) \tag{4}$$

Here, ΔG^* is the critical energy for nucleation, and is decreasing with increased degree of super-saturation [27]; ΔG_D is the critical energy for mass transport; and k_B is the Boltzmann constant.

The major differences between the AC and HP series are the synthesis method and composition. First, the HP samples are less brittle and are easier to handle during the subsequent measurements, with somewhat larger bulk density values compared to the AC ones, Table 2. Second, the HP series contains smaller concentrations of Ag (3.3 at. %) compared to AC (5.0 at. %). The reduction of Ag-concentration in the HP series enabled us avoiding formation of primary Ag_2Te-precipitates during quenching owing to reduced degree of super-saturation.

It is shown in Figure 3 that 6 h aging at 380 °C is the critical condition for obtaining an adequately high value of $N_V = 2.7 \cdot 10^{20}$ m^{-3}. Such value was found to be enough to initiate phonon scattering in PbTe [32,33]. After reaching at this peak N_V, an over-aging stage begins, in which the precipitates coarsen [27], and their effect on phonon scattering is expected to diminish.

Comparison between the microstructures obtained for the Bi-doped and Bi-free materials that underwent identical aging conditions shown in Figure 2—on the one hand, and the obtained N_V values shown in Figure 3—on the other hand, is very instructive. First, it is observed that Bi greatly suppresses temporal evolution of the precipitates. This is manifested by relatively large, needle-like precipitates, several micrometers long, which appear in the Bi-doped samples; this size is preserved for a wide range of aging times. Conversely, their undoped counterparts exhibit small precipitates (tens of nm dia.) for 6 h aging, and large precipitates (ca. 500 nm dia.) for 48 h aging. This implies that Bi atoms encourage "pinning" of the Ag_2Te-precipitates to their original sizes. Second, Bi additions affect the morphology of the precipitates. Whereas the undoped samples exhibit needle-shaped precipitates that evolve into spheroidal ones between 4 and 6 h aging, the Bi-doped samples contain only needle-shaped ones. This can be explained by coherent interfaces existing between the PbTe-matrix and Ag_2Te-precipitates due to small lattice mismatch [25]. The degree of coherency decreases during precipitate growth [27]; however, Bi-additions preserve high degree of coherency, possibly due to interfacial segregation, as observed in a different PbTe-based system [34].

4.2. Electronic Transport Properties

In this section we will explain the complicated behavior shown in Figure 4, starting from the Bi-free materials. Generally, all samples exhibit an inverse relation between σ and |S|, that is, as one increases the other one decreases with dopant concentration, which is expected [29]. Most samples, however, exhibit increase of σ and decrease of |S| with increasing temperature. This is opposite to the behavior observed for heavily-doped semiconductors [29], which is associated to the fact that charge carrier activation occurs from the Fermi level to higher energy states in the conduction band. Rather, these trends are often observed for intrinsic or lightly-doped semiconductors, where such behavior originates from the thermal activation of charge carriers across the band gap. This is, however, not the case here. Our explanation for this trend is based on the increase of Ag solubility limit in PbTe with increasing temperature, which leads to dissolution of Ag_2Te-precipitates and increase of the extrinsic charge carrier concentration with increasing temperature [24]; this raises the Fermi energy to levels closer to the conduction band.

It should be noted that Ag is an n-type dopant in Ag-saturated PbTe due to formation of interstitial Ag defects [24], therefore yields negative S-values, whereas for low concentrations Ag atoms act as acceptors. During aging, nucleation and growth of the Ag_2Te-phase take place, and Ag-atoms are consumed and depleted from the PbTe-matrix. Therefore, S-coefficients become less negative with increasing aging time, Figure 4b. Interestingly, the Bi-free sample aged for 48 h shows very low electrical conductivity, and its S-coefficient is positive and changes its sign to negative upon heating, implying a transition from p-type to n-type behavior. This can be explained considering that after 48 h aging the volume fraction of Ag_2Te is the greatest, so that the amount of Ag-atoms dissolved in the PbTe-matrix is the smallest. This is because Ag_2Te precipitates act as p-type dopants in PbTe, in analogy with Na_2Te and K_2Te, where Na^+ or K^+ substitute for Pb^{+2} [23]. Ca. 1% solubility of Ag_2Te in PbTe apparently results in compensated defects and very low concentration of extrinsic charge carriers ($<10^{18}$ cm^{-3}), possibly since half of the Ag atoms occupying interstitial sites donate one electron compensating for the remaining Ag substituting for Pb. The occurrence of Ag as both an n- and a p-type dopant has been reported previously [35]. Overall, Ag-solubility increases with temperature so that $|S|$ is decreasing with temperature for all cases. Moreover, Ag_2Te-precipitates dissolve at temperatures as high as 450 °C, which is manifested by convergence of all curves at this temperature for both Bi-alloyed and Bi-free compounds, Figure 4b.

The behavior of the Bi-alloyed compound is easier to comprehend, since Bi-dopants serve always as donors in PbTe. The as-quenched Bi-alloyed sample exhibits decreasing σ- values with growing temperatures starting from ca. 750 S·cm^{-1} at room temperature. This behavior can be explained as follows. At the beginning of the measurement cycle Bi atoms, acting as electron donors, are homogeneously dispersed in the matrix that contains very few Ag_2Te-precipitates per unit volume. During SBA analysis the temperature reaches 700 °C, followed by slow cooling. We hypothesize that Bi-atoms prefer to segregate to the interface between the PbTe-matrix and the Ag_2Te-precipitates to reduce the total interfacial free energy. Once Bi-atoms segregate to these interfaces, their concentration in the PbTe-matrix decrease. The electrical conductivity of the '0 h' sample is relatively high and keeps decreasing with temperature since the PbTe-matrix is sufficiently enriched by Bi. The samples aged for 6 h and more contain less Bi-atoms dissolved in the PbTe-matrix, so that their electrical conductivities are smaller than that of the '0 h' one, and they behave as in the non-degenerate limit, Figure 4a. This also explains why S-values become more negative with aging time. We note that this suggested mechanism relies on the scenario of interfacial segregation of Bi-atoms to the PbTe/Ag_2Te interfaces. To the best of our knowledge, this has not been reported in literature, however was validated by us from first-principles [36].

4.3. Thermal Transport Properties

Figure 5 shows an interesting behavior, in which the Bi-free compounds exhibit the lowest κ_l values for 6 h aging, whereas 48 h aging leads to maximum κ_l values. Conversely, the Bi-alloyed compounds exhibit κ_l values with increasing order of their aging times. This can be elucidated considering the role of Ag_2Te-precipitates in phonon scattering. The Bi-free samples exhibit decrease in κ_l-values with aging time from the as quenched state up to 6 h due to the significant increase of N_v to values as large as 2.7×10^{20} m^{-3}; such N_v-value was found to be adequately high to initiate phonon scattering with sufficient intensity to reduce thermal conductivity by tens percent [32]. Conversely,

κ_l-values increase from 6 h aging up to 48 h aging, which seems to deviate from the trend introduced before. This is explained by the role of the degree of matrix supersaturation in scattering phonons, as utilized by us previously [28,37]. The as-quenched sample comprises non-equilibrium, strained, and super-saturated PbTe solid solution that relaxes upon aging. For the as-quenched state, this is the dominating mechanism for phonon scattering. With growing aging time, the matrix strains relieve, however N_V increases. For longer aging times, N_V decreases due to coarsening with simultaneous strain relief; therefore, the lattice thermal conductivity grows to values even larger than for the as-quenched state. The interplay between these competing factors determine the temporal evolution of κ_l shown in Figure 5a. For comparison, the trends shown in Figure 5b for the Bi-alloyed samples are somewhat different: κ_l-values keep increasing with aging time. This is because the N_V-values of these samples are too low, so that precipitates do not play any significant role in phonon scattering. Instead, matrix relaxation remains the only mechanism, and this yields continuous increase of κ_l with aging time. To demonstrate the significant influence of precipitates having the same N_V-values as in Figure 3 on κ_l, we calculate $\kappa_l(T)$ for 10 nm radius precipitates dispersed in PbTe-matrix having different N_V-values applying the Callaway model with parameters that are conventional for PbTe-matrix [32,33], and the results are shown in Figure 6.

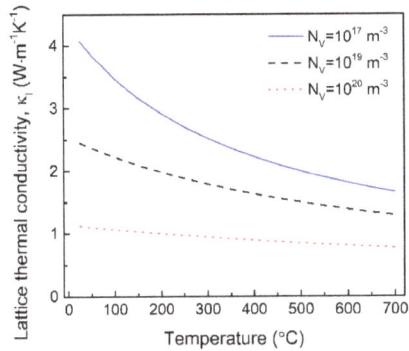

Figure 6. The temperature dependent lattice thermal conductivity values calculated based on the Callaway model for PbTe-matrix containing precipitates of 10 nm radius and three different number densities: $N_V = 10^{17}$, 10^{19}, and 10^{20} m^{-3}, denoted by blue solid, black dashed, and red dotted lines, respectively.

It is indicated that κ_l is sensitive to precipitate number density variations in the range simulated, which corroborates our analysis.

4.4. Thermoelectric Performance

We examine the effects of aging and microstructure evolution on the TE figure of merit, ZT. Figure 7 shows the temperature dependent ZT-values calculated for the HP series of both Bi-alloyed and Bi-free materials, together with selected values reported in literature for other n-type PbTe.

Interestingly, it is indicated that ZT is extremely sensitive to microstructure evolution for the Bi-free compounds, and is almost insensitive to it for the Bi-alloyed ones. Additionally, the Bi-free ones exhibit maximum ZT-values that are greater than those of the Bi-alloyed ones. Most importantly, ZT significantly improves due to 6 h aging with respect to the raw material (as-quenched), which is associated to the decrease of κ_l thanks to precipitation of the Ag$_2$Te-phase. The maximum ZT value obtained is ca. 1.4 at 450 °C, which is one of the largest ones reported for n-type PbTe [38–43]. Subsequent aging for 48 h results in drastic decrease of ZT for the undoped specimens, whereas that of the Bi-doped specimen only slightly decreases. This can be explained by coarsening process taking place in the undoped specimen after 48 h aging, Figures 2 and 3, which increases lattice thermal conductivity and reduces electrical conductivity. Furthermore, Bi alloying suppresses the microstructure evolution and reduces the N_V-values, Figure 2, which leads to only a slight reduction of ZT after 6 and 48 h aging. This clarifies the positive effects of Bi in stabilizing the material's microstructure against evolution, and thereby in preventing deterioration of TE performance. All of the above imply that good control and understanding of the evolved microstructure help us to employ two-phase materials for TE applications.

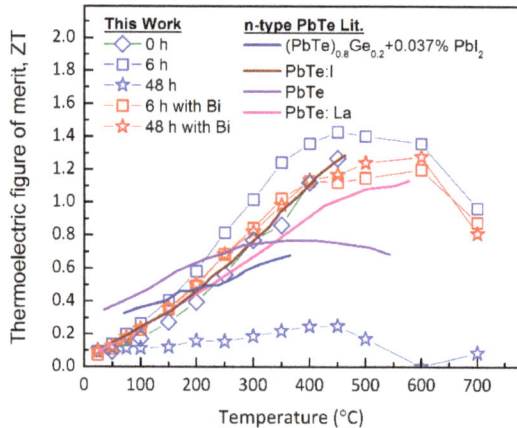

Figure 7. The temperature dependent thermoelectric figure of merit, ZT, evaluated for the hot-pressed (HP) series of specimens aged at 380 °C for 0 (diamonds), 6 (squares), and 48 h (stars) for the Bi-free (blue symbols) and Bi-alloyed (red symbols) materials. Selected data for n-type PbTe from literature are plotted for comparison.

5. Summary and Conclusions

Our investigation of Ag_2Te precipitation in undoped and Bi-doped $(PbTe)_{1-x}(Ag_2Te)_x$ compounds includes materials synthesis in two different routes, namely casting and hot-pressing, aging heat treatments, microstructure characterization, and TE property measurements. It was found that Ag_2Te-precipitate number density can be controlled by changing the average composition as well as heat treatment temperature and duration. A maximum precipitate number density value as high as 2.7×10^{20} m^{-3} was achieved for the Bi-free compounds after 6 h aging at 380 °C, yielding ZT = 1.4 at 450 °C. This is one on the largest values reported for n-type PbTe compounds, and is associated to improved phonon scattering efficiency. Subsequent aging results in precipitate coarsening, which causes drastic increase of thermal conductivity and decrease of electrical conductivity. This is associated to a reduction of phonon scattering efficiency by both precipitates and matrix strains, as well as depletion of Ag-solutes from the PbTe-matrix. Interestingly, Bi-doping results in stagnation of microstructure evolution, maintaining thermally stable TE performance. Overall, this study provides us with fundamental understanding and practical tools necessary to design TE properties in PbTe-based compounds as well as in other two-phase systems.

Acknowledgments: This research was carried out in the framework of the Nancy & Stephen Grand Technion Energy Program (GTEP), and supported by the Leona M. and Harry B. Helmsley Charitable Trust and the Adelis Foundation for renewable energy research. Generous support from the Israel Science Foundation (ISF), Grant No. 698/13, as well as from the German-Israeli Foundation for Research and Development (GIF), Grant No. I-2333-1150.10/2012 is gratefully acknowledged.

Author Contributions: T.G. performed materials synthesis, microstructure characterization and thermoelectric measurements, and analyzed the data; Y.A. and Y.G. designed the experimental procedures and methodology, and provided scientific supervision; Y.G. supervised hot-pressing processes; A.S. contributed to materials characterization and data analysis; T.G. and Y.A. wrote the paper. All authors contributed to this research.

Crystals **2017**, *7*, 281

Conflicts of Interest: The authors declare no conflict of interest.

References

1. LaLonde, A.D.; Pei, Y.; Wang, H.; Snyder, G.J. Lead telluride alloy thermoelectrics. *Mater. Today* **2011**, *14*, 526–532. [CrossRef]
2. Snyder, G.J.; Toberer, E.S. Complex thermoelectric materials. *Nat. Mater.* **2008**, *7*, 105–114. [CrossRef] [PubMed]
3. Dughaish, Z.H. Lead telluride as a thermoelectric material for thermoelectric power generation. *Phys. B Condens. Matter* **2002**, *322*, 205–223. [CrossRef]
4. Tan, G.; Shi, F.; Hao, S.; Zhao, L.D.; Chi, H.; Zhang, X.; Uher, C.; Wolverton, C.; Dravid, V.P.; Kanatzidis, M.G. Non-equilibrium processing leads to record high thermoelectric figure of merit in PbTe-SrTe. *Nat. Commun.* **2016**, *7*, 12167. [CrossRef] [PubMed]
5. Korkosz, R.J.; Chasapis, T.C.; Lo, S.h.; Doak, J.W.; Kim, Y.J.; Wu, C.I.; Hatzikraniotis, E.; Hogan, T.P.; Seidman, D.N.; Wolverton, C.; et al. High ZT in p-type $(PbTe)_{1-2x}(PbSe)_x(PbS)_x$ thermoelectric materials. *J. Am. Chem. Soc.* **2014**, *136*, 3225–3237. [CrossRef] [PubMed]
6. Rawat, P.K.; Paul, B.; Banerji, P. Exploration of Zn resonance levels and thermoelectric properties in I-doped PbTe with ZnTe nanostructures. *ACS Appl. Mater. Interfaces* **2014**, *6*, 3995–4004. [CrossRef] [PubMed]
7. Tan, G.; Kanatzidis, M.G. Chapter 4 All-Scale Hierarchical PbTe. In *Materials Aspect of Thermoelectricity*; CRC Press: Boca Raton, FL, USA, 2016; pp. 125–158.
8. Tritt, T.M. *Thermal Conductivity: Theory, Properties, and Applications*; Springer: Berlin, Germany, 2004.
9. Rowe, D.M. *Thermoelectrics Handbook: Macro to Nano*; CRC Press: Boca Raton, FL, USA, 2006.
10. Wu, D.; Zhao, L.-D.; Zheng, F.; Jin, L.; Kanatzidis, M.G.; He, J. Understanding nanostructuring processes in thermoelectrics and their effects on lattice thermal conductivity. *Adv. Mater.* **2016**, *28*, 2737–2743. [CrossRef] [PubMed]
11. Li, J.; Zhang, X.; Lin, S.; Chen, Z.; Pei, Y. Realizing the high thermoelectric performance of GeTe by Sb-doping and Se-alloying. *Chem. Mater.* **2017**, *29*, 605–611. [CrossRef]
12. He, J.Q.; Sootsman, J.R.; Girard, S.N.; Zheng, J.C.; Wen, J.G.; Zhu, Y.M.; Kanatzidis, M.G.; Dravid, V.P. On the origin of increased phonon scattering in nanostructured PbTe based thermoelectric materials. *J. Am. Chem. Soc.* **2010**, *132*, 8669–8675. [CrossRef] [PubMed]
13. Medlin, D.L.; Snyder, G.J. Interfaces in bulk thermoelectric materials a review for current opinion in colloid and interface science. *Curr. Opin. Colloid Interface Sci.* **2009**, *14*, 226–235. [CrossRef]
14. Ikeda, T.; Collins, L.A.; Ravi, V.A.; Gascoin, F.S.; Haile, S.M.; Snyder, G.J. Self-assembled nanometer lamellae of thermoelectric PbTe and Sb_2Te_3 with epitaxy-like interfaces. *Chem. Mater.* **2007**, *19*, 763–767. [CrossRef]
15. Ikeda, T.; Haile, S.M.; Ravi, V.A.; Azizgolshani, H.; Gascoin, F.; Snyder, G.J. Solidification processing of alloys in the pseudo-binary PbTe- Sb_2Te_3 system. *Acta Mater.* **2007**, *55*, 1227–1239. [CrossRef]
16. Ikeda, T.; Iwanaga, S.; Wu, H.J.; Marolf, N.J.; Chen, S.W.; Snyder, G.J. A combinatorial approach to microstructure and thermopower of bulk thermoelectric materials: The pseudo-ternary PbTe-Ag_2Te-Sb_2Te_3 system. *J. Mater. Chem.* **2012**, 24335–24347. [CrossRef]
17. Ikeda, T.; Marolf, N.J.; Bergum, K.; Toussaint, M.B.; Heinz, N.A.; Ravi, V.A.; Jeffrey Snyder, G. Size control of Sb_2Te_3 Widmanstätten precipitates in thermoelectric PbTe. *Acta Mater.* **2011**, *59*, 2679–2692. [CrossRef]
18. Ikeda, T.; Ravi, V.; Jeffrey Snyder, G. Microstructure size control through cooling rate in thermoelectric PbTe-Sb_2Te_3 composites. *Metall. Mater. Trans. A* **2010**, *41*, 641–650. [CrossRef]
19. Ikeda, T.; Ravi, V.A.; Snyder, G.J. Formation of Sb_2Te_3 Widmanstätten precipitates in thermoelectric PbTe. *Acta Mater.* **2009**, *57*, 666–672. [CrossRef]
20. Ikeda, T.; Toberer, E.S.; Ravi, V.A.; Snyder, G.J.; Aoyagi, S.; Nishibori, E.; Sakata, M. In situ observation of eutectoid reaction forming a PbTe-Sb_2Te_3 thermoelectric nanocomposite by synchrotron X-ray diffraction. *Scr. Mater.* **2009**, *60*, 321–324. [CrossRef]
21. Ikeda, T.; Toussaint, M.; Bergum, K.; Iwanaga, S.; Jeffrey Snyder, G. Solubility and formation of ternary Widmanstätten precipitates in PbTe in the pseudo-binary PbTe–Bi_2Te_3 system. *J. Mater. Sci.* **2011**, *46*, 3846–3854. [CrossRef]

22. Pei, Y.; Heinz, N.A.; LaLonde, A.; Snyder, G.J. Combination of large nanostructures and complex band structure for high performance thermoelectric lead telluride. *Energy Environ. Sci.* **2011**, *4*, 3640–3645. [CrossRef]

23. Pei, Y.; Lensch-Falk, J.; Toberer, E.S.; Medlin, D.L.; Snyder, G.J. High thermoelectric performance in PbTe due to large nanoscale Ag_2Te precipitates and La doping. *Adv. Funct. Mater.* **2011**, *21*, 241–249. [CrossRef]

24. Pei, Y.; May, A.F.; Snyder, G.J. Self-tuning the carrier concentration of $PbTe/Ag_2Te$ composites with excess Ag for high thermoelectric performance. *Adv. Energy Mater.* **2011**, *1*, 291–296. [CrossRef]

25. Lensch-Falk, J.L.; Sugar, J.D.; Hekmaty, M.A.; Medlin, D.L. Morphological evolution of Ag_2Te precipitates in thermoelectric PbTe. *J. Alloys Compd.* **2010**, *504*, 37–44. [CrossRef]

26. Bergum, K.; Ikeda, T.; Jeffrey Snyder, G. Solubility and microstructure in the pseudo-binary $PbTe-Ag_2Te$ system. *J. Solid State Chem.* **2011**, *184*, 2543–2552. [CrossRef]

27. Porter, D.A.; Easterling, K.E. *Phase Transformations in Metals and Alloys*, 2nd ed.; Chapman & Hall: London, UK, 1992.

28. Koresh, I.; Amouyal, Y. Effects of microstructure evolution on transport properties of thermoelectric nickel-doped zinc oxide. *J. Eur. Ceram. Soc.* **2017**, *37*, 3541–3550. [CrossRef]

29. Goldsmid, H.J. *Introduction to Thermoelectricity*; Springer: Berlin/Heidelberg, Germany, 2009.

30. Kim, H.-S.; Gibbs, Z.M.; Tang, Y.; Wang, H.; Snyder, G.J. Characterization of Lorenz number with Seebeck coefficient measurement. *APL Mater.* **2015**, *3*, 041506. [CrossRef]

31. Pei, Y.-L.; Liu, Y. Electrical and thermal transport properties of Pb-based chalcogenides: PbTe, PbSe, and PbS. *J. Alloys Compd.* **2012**, *514*, 40–44. [CrossRef]

32. Amouyal, Y. Reducing lattice thermal conductivity of the thermoelectric compound $AgSbTe_2$ (P4/mmm) by lanthanum substitution: Computational and experimental approaches. *J. Electron. Mater.* **2014**, *43*, 3772–3779. [CrossRef]

33. Amouyal, Y. A practical approach to evaluate lattice thermal conductivity in two-phase thermoelectric alloys for energy applications. *Materials* **2017**, *10*, 386. [CrossRef] [PubMed]

34. He, J.; Blum, I.D.; Wang, H.Q.; Girard, S.N.; Doak, J.; Zhao, L.D.; Zheng, J.C.; Casillas, G.; Wolverton, C.; Jose-Yacaman, M.; et al. Morphology control of nanostructures: Na-doped PbTe–PbS system. *Nano Lett.* **2012**, *12*, 5979–5984. [CrossRef] [PubMed]

35. Strauss, A.J. Effect of Pb- and Te-saturation on carrier concentrations in impurity-doped PbTe. *J. Electron. Mater.* **1973**, *2*, 553–569. [CrossRef]

36. Grossfeld, T. Microstructure Evolution and Enhancement of the Thermoelectric Conversion Efficiency of PbTe-Based Compounds for Renewable Energy Applications. Mater's Thesis, Technion—Israel Institute of Technology, Haifa, Israel, 2015.

37. Cojocaru-Mirédin, O.; Abdellaoui, L.; Nagli, M.; Zhang, S.; Yu, Y.; Scheu, C.; Raabe, D.; Wuttig, M.; Amouyal, Y. Role of nanostructuring and microstructuring in silver antimony telluride compounds for thermoelectric applications. *ACS Appl. Mater. Interfaces* **2017**, *9*, 14779–14790. [CrossRef] [PubMed]

38. Pei, Y.; LaLonde, A.; Iwanaga, S.; Snyder, G.J. High thermoelectric figure of merit in heavy hole dominated PbTe. *Energy Environ. Sci.* **2011**, *4*, 2085–2089. [CrossRef]

39. Pei, Y.; Wang, H.; Snyder, G.J. Band engineering of thermoelectric materials. *Adv. Mater.* **2012**, *24*, 6125–6135. [CrossRef] [PubMed]

40. Pei, Y.; Shi, X.; LaLonde, A.; Wang, H.; Chen, L.; Snyder, G.J. Convergence of electronic bands for high performance bulk thermoelectrics. *Nature* **2011**, *473*, 66–69. [CrossRef] [PubMed]

41. Wang, H.; Pei, Y.; LaLonde, A.D.; Snyder, G.J. Weak electron–phonon coupling contributing to high thermoelectric performance in n-type PbSe. *Proc. Natl. Acad. Sci. USA* **2012**, *109*, 9705–9709. [CrossRef] [PubMed]

42. Wang, H.; Pei, Y.; LaLonde, A.D.; Snyder, G.J. Heavily doped p-type PbSe with high thermoelectric performance: An alternative for PbTe. *Adv. Mater.* **2011**, *23*, 1366–1370. [CrossRef] [PubMed]

43. Pei, Y.; LaLonde, A.D.; Wang, H.; Snyder, G.J. Low effective mass leading to high thermoelectric performance. *Energy Environ. Sci.* **2012**, *5*, 7963–7969. [CrossRef]

© 2017 by the authors. Licensee MDPI, Basel, Switzerland. This article is an open access article distributed under the terms and conditions of the Creative Commons Attribution (CC BY) license (http://creativecommons.org/licenses/by/4.0/).

crystals

MDPI

Article

Spark Plasma Sintering of Tungsten Oxides WO$_x$ (2.50 ≤ x ≤ 3): Phase Analysis and Thermoelectric Properties

Felix Kaiser [1,*], Paul Simon [1], Ulrich Burkhardt [1], Bernd Kieback [2], Yuri Grin [1] and Igor Veremchuk [1,*]

[1] Max-Planck-Institut für Chemische Physik fester Stoffe, 01187 Dresden, Germany; paul.simon@cpfs.mpg.de (P.S.); burkhard@cpfs.mpg.de (U.B.); grin@cpfs.mpg.de (Y.G.)
[2] Fraunhofer Institut für Fertigungstechnik und Angewandte Materialforschung, 01277 Dresden, Germany; bernd.kieback@ifam-dd.fraunhofer.de
* Correspondence: felix.kaiser@cpfs.mpg.de (F.K.); igor.veremchuk@cpfs.mpg.de (I.V.); Tel.: +49-351-4646-4000 (F.K. & I.V.)

Academic Editor: George S. Nolas
Received: 26 July 2017; Accepted: 23 August 2017; Published: 5 September 2017

Abstract: The solid-state reaction of WO$_3$ with W was studied in order to clarify the phase formation in the binary system W-O around the composition WO$_x$ (2.50 ≤ x ≤ 3) during spark plasma sintering (SPS). A new phase "WO$_{2.82}$" is observed in the range 2.72 ≤ x ≤ 2.90 which might have the composition W$_{12}$O$_{34}$. The influence of the composition on the thermoelectric properties was investigated for 2.72 ≤ x ≤ 3. The Seebeck coefficient, electrical conductivity and electronic thermal conductivity are continuously tunable with the oxygen-to-tungsten ratio. The phase formation mainly affects the lattice thermal conductivity κ_{lat} which is significantly reduced until 700 K for the sample with the composition x = 2.84, which contains the phases W$_{18}$O$_{49}$ and "WO$_{2.82}$". In single-phase WO$_{2.90}$ and multi-phase WO$_x$ materials (2.90 ≤ x ≤ 3), which contain crystallographic shear plane phases, a similar reduced κ_{lat} is observed only below 560 K and 550 K, respectively. Therefore, the composition range x < 2.90 in which the pentagonal column structural motif is formed might be more suitable for decreasing the lattice thermal conductivity at high temperatures.

Keywords: thermoelectric materials; spark plasma sintering; tungsten oxides; crystallographic shear plane phases

1. Introduction

Transition metal oxides (TMOs) are under debate for high-temperature thermoelectric (TE) applications since NaCo$_2$O$_4$ was found to be a p-type material with a thermopower α as high as 100 μV·K^{-1} at 300 K [1]. By now, TMOs are subject of in-depth investigations [2–4] in particular because there is still a lack of appropriate n-type counterparts, although lanthanum-doped SrTiO$_3$ [5], ZnO [6] or doped TiO$_{2-x}$ [7–9] show promising results.

The enhancement of a TE material's figure-of-merit ZT (Equation (6)) might be obtained by either increasing the power factor $\alpha^2\sigma$ or decreasing the total thermal conductivity $\kappa_{tot} = \kappa_{el} + \kappa_{lat}$ (Equations (3) and (4)). An interesting approach is the exploitation of crystallographic shear (CS) for the reduction of the lattice thermal conductivity κ_{lat} due to increased phonon scattering [4,10,11]. CS plane structures occur when a transition metal is partially reduced and oxygen layers are removed from the structure, e.g., of TiO$_2$, VO$_2$, MoO$_3$ and WO$_3$. In the tungsten-oxygen system (Figure 1a) with decreasing O/W-ratio x the observed phases are the insulating ReO$_3$-type WO$_3$, the semiconducting phase W$_{20}$O$_{58}$ with CS planes, the metallic phase W$_{18}$O$_{49}$ with pentagonal columns (PC) and the metallic rutile-type WO_2, with varying coupling of the $[WO_6]$ octahedra (Figure 1b–e). Such structure modifications result in an increasing charge carrier concentration [12]. A further five observed phases in the range $2 < x < 3$ are predicted to be bad metals with charge carrier densities of the order 10^{21}–10^{22} cm^{-1} according to band structure calculations [13]. Yet, these phases occur only in small domains of single crystals [14,15] or under high pressure conditions [16–18]. The fact that they are not observed as bulk materials might be attributed to a microstrain-driven ordering mechanism of the CS planes [19], which should be sensitive to the synthesis conditions.

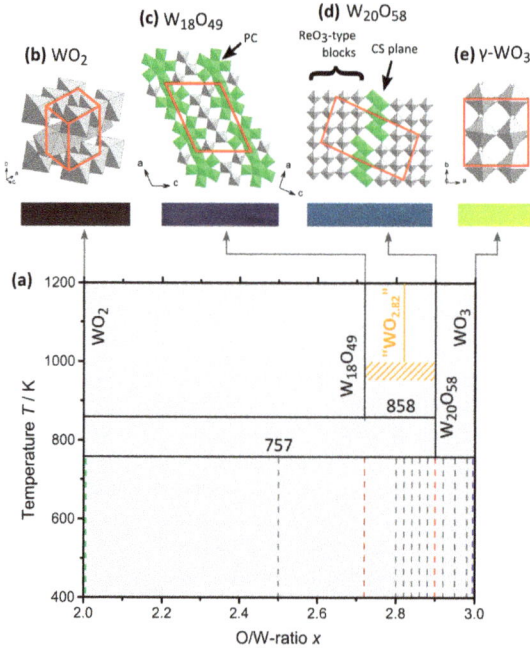

Figure 1. (**a**) Phase diagram of the binary system W-O in the range of the O/W-ratio $2 \leq x \leq 3$ [20]. Dashed lines mark the sample compositions prepared and characterized in this work. A metastable phase "$WO_{2.82}$" (orange) is found; (**b–e**) Crystal structures and respective colors of the known tungsten oxide phases. Characteristic features are the $[WO_6]$ octahedra (grey), pentagonal columns (PC) and crystallographic shear (CS) planes (green). For clarity, only the monoclinic γ-WO_3 is shown among the WO_3 modifications.

The vapor-transport preparation route to obtain these phases comprises a high-temperature heating of WO_x powder mixtures over several days and weeks [14,15,21–23]. Recently $WO_{2.72}$ ($W_{18}O_{49}$) and $WO_{2.90}$ ($W_{20}O_{58}$) were successfully prepared by spark plasma sintering (SPS) being promising bulk TE materials [11,12,24]. SPS combines the solid-state reaction of powdered precursor mixtures with simultaneous shaping, and provides an effective manufacturing route for TE materials due to low temperatures and short reaction times [8]. The samples around the composition $WO_{2.90}$ still showed the formation of a further phase which was interpreted as $WO_{2.96}$ [25].

The knowledge on phase formation in the tungsten-oxide system under SPS conditions still appears fragmentary which complicates the evaluation of the system's potential as a TE material. Thus, we studied the products of the SPS redox reaction dependent on both, the composition x in the range $2.50 \leq x \leq 3$ (Figure 1a, dashed lines) and the synthesis temperature T_{max} with powder X-ray diffraction (PXRD). Subsequently, the TE properties of the SPS-prepared WO_x materials with $x \geq 2.72$ were investigated.

In this work we use the notation "WO$_x$" for sample compositions, and "W$_a$O$_b$" for phases and their crystal structures.

2. Results & Discussion

2.1. Spark Plasma Preparation

The solid-state reaction

$$x/3\ WO_3 + (1 - x/3)\ W \rightarrow WO_x \tag{1}$$

was performed for different x by spark plasma sintering (SPS) in vacuum (<10 Pa) under an uniaxial pressure of 80 MPa (Figure 2a). Temperature programs included a linear heating with 50 K·min^{-1}, an isothermal dwell time t_{dwell} at the temperature T_{max}, and free cooling. Specific experiments in this work are notated by "SPS-T_{max}/t_{dwell}".

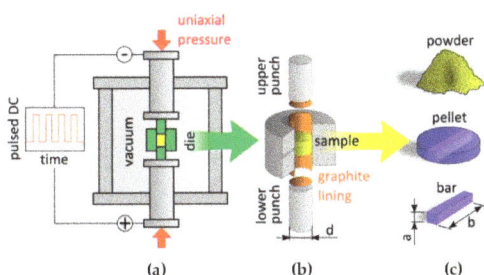

Figure 2. (**a**) Setup of the spark plasma sintering (SPS) machine; (**b**) Scheme of the filled SPS die; (**c**) Sample powder reacted in the SPS and compacted to pellets. For measurements tetragonal bars with a ≈ 1.5 mm and b ≈ 6–8 mm were cut.

For the reference material WO$_{2.90}$, the SPS treatment was optimized regarding the phase purity. Systematic variations of the temperature and dwell time in the ranges 1320 K $\leq T_{max} \leq$ 1570 K and 10 min $\leq t_{dwell} \leq$ 4 h, respectively, yielded the optimum regime SPS-1420 K/10 min (Figure 3b). Samples WO$_x$ with x = {2.50, 2.72, 2.80, 2.82, 2.84, 2.86, 2.88, 2.92, 2.95, 2.98} were prepared with the same conditions.

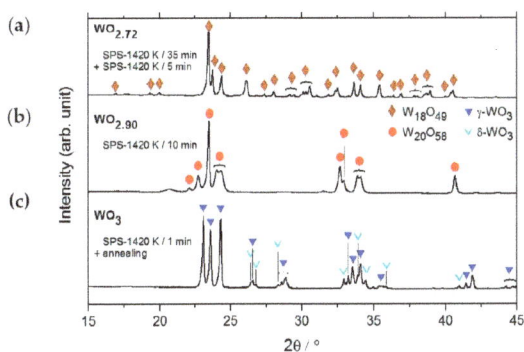

Figure 3. PXRD patterns (Cu-K$_{\alpha 1}$ radiation) of the reference materials obtained by SPS: (**a**) Single-phase WO$_{2.72}$ after two-step treatment with surface cleaning and grinding in between; (**b**) Single-phase WO$_{2.90}$ after one-step treatment; (**c**) WO$_3$ sample contains two modification γ- and δ-WO$_3$ after SPS and subsequent annealing in open air (1170 K, 68 h).

Phase-pure reference material $WO_{2.72}$ was obtained via SPS-1420 K/35 min followed by SPS-1420 K/5 min with cleaning of the surface and grinding of the sample between both steps (Figure 2a).

A WO_2 reference sample was directly compacted from the as-purchased material with SPS-1470 K/10 min (green curve in Figure 4).

The compaction of the yellowish as-purchased material WO_3 with SPS-1420 K/10 min resulted in a bluish-violet and multiple-phase product according to PXRD (blue curve in Figure 4). Subsequent annealing performed at 1170 K for 68 h in the open air was supposed to yield the monoclinic γ-WO_3 [26] and resulted in a greyish-yellow specimen (Figure 3c).

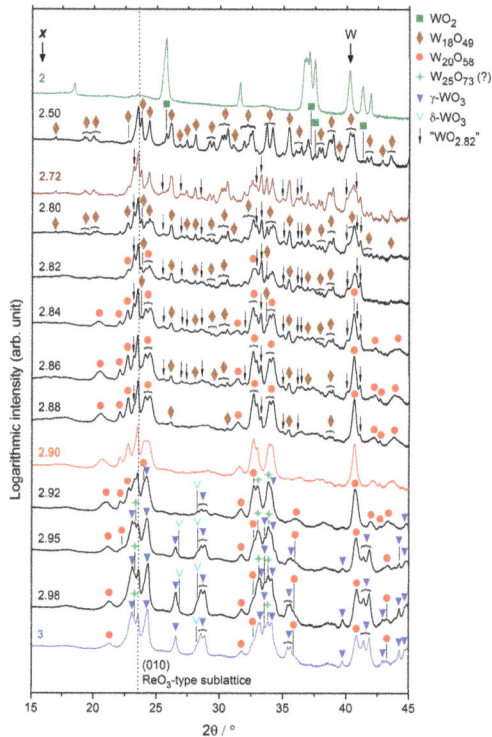

Figure 4. PXRD patterns (Cu-K$_{\alpha1}$ radiation) of the WO_x samples processed with SPS-1420/10 min. Only the WO_2 reference sample was compacted with SPS-1470 K/10 min (green line).

2.2. Single-Phase Materials WO_2, $WO_{2.72}$, $WO_{2.90}$ and WO_3

According to PXRD, phase-pure material $WO_{2.72}$ is obtained from the two-step treatment (Figure 3a).

Phase-pure material $WO_{2.90}$ is yielded from the routine SPS-1420 K/10 min as seen from the PXRD pattern (Figure 3b), which shows only reflections of $W_{20}O_{58}$ ($x = 2.90$). However, all reflections except $(0\,k\,0)$ show large FWHM values indicating good ordering only in the [010] direction (Figure 1d). Starting from this optimized regime for $x = 2.90$, the increase of both T_{max} and t_{dwell} promotes the formation of $W_{18}O_{49}$ ($x = 2.72$), WO_2 and an additional phase. We ascribe the composition $x \approx 2.82$ for this phase as seen in the following. Thus, the schematic secondary reaction

$$W_{20}O_{58} \rightarrow W_{18}O_{49} + WO_2 + \text{``}WO_{2.82}\text{''} \tag{2}$$

represents a further tungsten reduction below $x = 2.90$ and is probably a similar SPS-specific reduction effect which is observed for pure WO_3. However, in the case of $W_{20}O_{58}$ a further decomposition can be kinetically inhibited with a short SPS treatment time $t_{dwell} = 10$ min.

The WO_2 does not undergo changes during the SPS treatment. A minor amount of elemental tungsten is observed in both, the as-purchased and SPS-processed material (green curve in Figure 4).

The as-purchased WO_3 sample is confirmed to adopt monoclinic structure (γ-WO_3). In contrast, the PXRD pattern after SPS treatment (SPS-1420 K/10 min, blue curve in Figure 4) shows a mixture of γ-WO_3 (blue triangles), triclinic δ-WO_3 (cyan arrows), $W_{20}O_{58}$ ($x = 2.90$, red dots), and an additional phase (green stars) whose reflections resemble those of $W_{25}O_{73}$ ($x = 2.92$) [14]. The δ-WO_3 is the stable modification below 290 K [27]. Conventional preparation of WO_3 via high-temperature oxidation usually yields γ-WO_3 [28] whereas the conditions during the free cooling in the SPS processing could promote the further transition $\gamma \rightarrow \delta$. After annealing in open air (1170 K, 68 h), only γ- and δ-WO_3 is found in the reference sample (Figure 3c). Most possibly, the annealing time was not sufficient to complete the $\delta \rightarrow \gamma$ transformation, or the cooling rate afterwards was not sufficient to suppress the $\gamma \rightarrow \delta$ transition completely. No clear evidence is found for the existence of $W_{25}O_{73}$ due to strong reflection overlapping. Yet, a reducing influence of the SPS graphite die and lining on the WO_3 starting material is clearly seen from the formation of $W_{20}O_{58}$. It is supposed to appear at ≈ 1020 K [29] and occurs despite the short $t_{dwell} = 10$ min and even 1 min. This SPS specific reduction should be considered additionally to the reaction with W, and will result in WO_x samples with compositions slightly deviating from the nominal O/W-ratio x.

The refined lattice parameters of the as-purchased materials and obtained single-phase products are given in Table 1 together with the respective literature data.

Table 1. Lattice parameters of phases in the tungsten–oxygen binary system refined from PXRD patterns in this work in comparison to literature data (iv—long-term heating in vacuo, SPS—spark plasma sintering, n.a.—not specified in the reference).

x	Phase	Source	Space Group	Lattice Parameters					Ref.
				a	b	c	β	V	
				Å	Å	Å	deg	Å^3	
2	WO_2	n.a.	$P2_1/c$	5.58	4.90	5.664	120.7	133.1	[30]
		commercial		5.574(1)	4.898(1)	5.662(1)	120.69(1)	132.9(2)	this work
		SPS		5.575(1)	4.900(1)	5.663(1)	120.70(1)	133.0(1)	this work
2.72	$W_{18}O_{49}$	n.a.	$P2/m$	18.32	3.79	14.04	115.0	883.3	[31]
		iv		18.33	3.79	14.04	115.2	882.1	[32]
		n.a.		18.32	3.78	14.03	115.2	879.5	[33]
		SPS		18.32	3.79	14.04	n/a	—	[11]
		SPS		18.329(1)	3.784(1)	14.037(1)	115.20(1)	880.9(4)	this work
2.83	$W_{12}O_{34}$	n.a.	$P2/m$	17.0	3.8	19.4	105.3		[34]
		SPS [a]		17.229(1)	3.782(1)	19.496(1)	105.77(1)	1223(3)	this work
2.90	$W_{20}O_{58}$	iv	$P2/m$	12.1	3.78	23.4	85	1066.2	[21]
		calculated		12.05	3.77	23.59	85.3	1067.2	[35]
		SPS		12.08	3.78	23.59	n/a	—	[11]
		SPS		12.00	3.78	23.51	84.8	1062.0	[24]
		SPS		12.080(3)	3.782(1)	23.62(1)	85.36(1)	1075.6(7)	this work
2.92	$W_{25}O_{73}$	iv	$P2/c$	11.93	3.82	59.72	98.3	2693.1	[14]
3	γ-WO_3	commercial	$P2_1/n$	7.33	7.56	7.73	90.5	428.3	[36]
		commercial		7.299(1)	7.537(1)	7.689(1)	90.88(1)	422.9(1)	this work

[a] Parameters determined from only 10 main reflections.

2.2.1. Crystal Structure on Atomic Resolution

From high-resolution transmission electron microscopy (HR-TEM) along [001] of the single-phase $WO_{2.90}$, large ordering areas are observed (Figure 5a). The zoomed and Fourier-filtered image from the top-left area (Figure 5b), and the corresponding fast Fourier transform (FFT) reveal the expected values of the lattice parameters a and b, but also some superstructure reflections indicating the doubling of the lattice parameter c (Figure 5c). In [010] direction, CS planes are visible but massively disordered with varying alignments and spacing (Figure 5d) which are consistent with the large FWHM values in the PXRD pattern (Figure 4a). The zoomed and filtered image (Figure 5e) shows low ordering along [001], where only the (001) reflection is present, whereas along [100], reflections of the fifth- and even higher order are observed in the FFT image (Figure 5f).

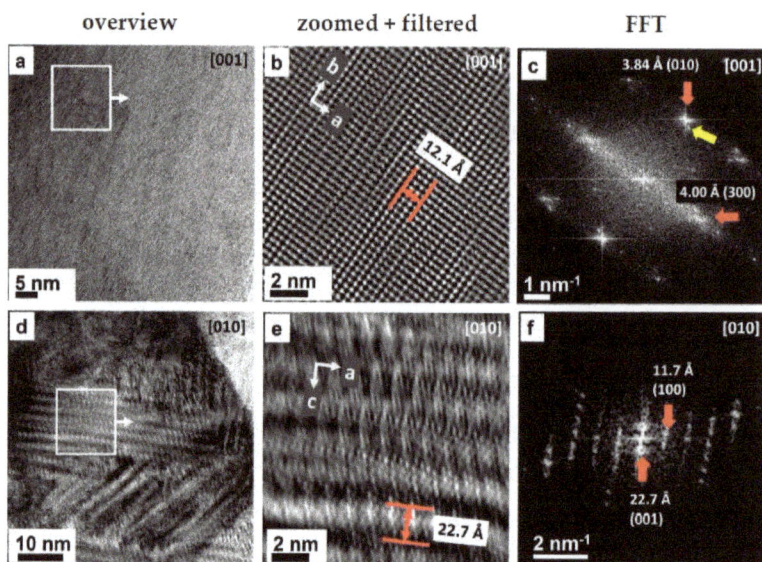

Figure 5. HR-TEM images of SPS-prepared $WO_{2.90}$ material along (**a–c**) [001] and (**d–f**) [010]. (**a**) Good ordering is found for the zone [001]; (**b**) The Fourier-filtered micrograph reveals a lattice with long-range variations; (**c**) Yellow-marked reflections in the corresponding FFT image indicate a superstructure; (**d**) CS planes are found for the [010] zone, however massive disorder is observed with deviations of CS plane orientations of about 44°; (**e**) Zooming and filtering reveals short-range order along [001]; (**f**) FFT with broadened and diffuse (001) reflection confirms disturbance of the atomic arrangement along [001].

2.2.2. Thermoelectric Properties

Among the known phases, the lowest electrical conductivity $\sigma(T)$ values are measured for $WO_{2.90}$ with $77(3) \times 10^3$ $S \cdot m^{-1}$ over the whole temperature range, which is on the level of a heavily doped semiconductor (Figure S1a). A local maximum at 635 K is in good accordance with previous records [11,12,24]. The electrical conductivity of WO_2 and $WO_{2.72}$ indicates typical metallic behaviour with $\sigma(T) \propto 1/T$ dependency (Figure S1b). No electrical transport measurement was possible for the compacted and annealed WO_3 sample due to its insulating behaviour. This indicates the absence of oxygen vacancies after SPS and annealing, which would promote the electrical conductivity [37].

Both metallic samples (WO_2 and $WO_{2.72}$) reveal a relative high thermal conductivity $\kappa_{tot}(T) > 10$ $W \cdot m^{-1} \cdot K^{-1}$ (Figure 6a). However, in the dense WO_2 structure (Figure 1b), where strong tungsten–tungsten interactions are expected, this arises mainly from the lattice contribution κ_{lat}, whereas in $WO_{2.72}$ κ_{el} and κ_{lat} contribute equally (Figure 6b). For the electrically insulating WO_3 sample ($\kappa_{lat} = \kappa_{tot}$), the temperature dependence $\kappa_{lat}(T)$ appears similar to that of $WO_{2.72}$. However, for $WO_{2.72}$ the approximation of κ_{lat} shows uncertainty of maximum 45% and a significant difference regarding WO_3 cannot be determined.

WO$_{2.90}$ shows $\kappa_{tot} \approx 3.5$–4.5 W·m^{-1}·K^{-1} over the whole temperature range (Figure 6a), which is close to previously published values [11,24]. With respect to errors its lattice contribution κ_{lat} equals that of WO$_3$ for $T > 560$ K (Figure 6b). Thus, the effect of CS planes in the W$_{20}$O$_{58}$ structure on the lattice thermal conductivity appears insignificant at high temperatures. The pentagonal column (PC) structural motif of the W$_{18}$O$_{49}$ phase (Figure 1c) in WO$_{2.72}$ might be an alternative phonon scattering center but proof requires a determination method for κ_{lat} which is more precise than our approximation from the Wiedemann-Franz law.

A direct property comparison of WO$_2$, WO$_{2.72}$, WO$_{2.90}$ and WO$_3$ can be conducted up to the maximum temperature $T = 863$ K at which WO$_{2.72}$ was investigated. Here WO$_{2.90}$ exhibits the highest Seebeck coefficient with $\alpha = -55$ µV·K^{-1} in contrast to WO$_{2.72}$ and WO$_2$ with -24 µV·K^{-1} and -22 µV·K^{-1} respectively (Figure S2a).

The figures of merit ZT at $T = 860$ K are low (0.006, 0.017 and 0.045 for WO$_2$, WO$_{2.72}$ and WO$_{2.90}$, respectively, Figure S2b). Maximum $ZT = 0.061$ is found for WO$_{2.90}$ at 963 K. The low Seebeck coefficient is a drawback for WO$_{2.72}$ and WO$_2$, together with the high κ_{lat} of WO$_2$. However, in the measured temperature range none of the materials shows a ZT maximum as of yet.

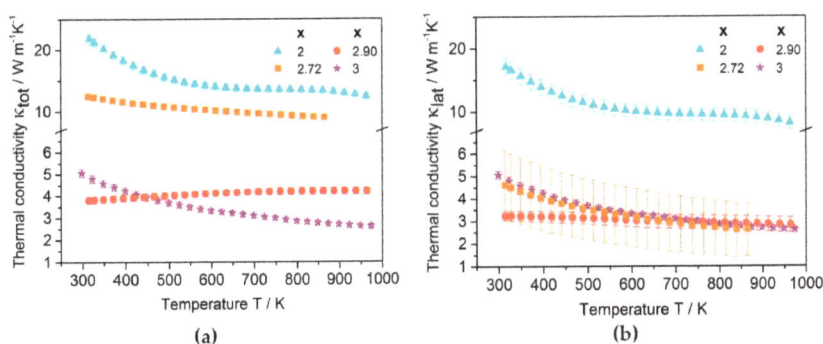

Figure 6. Thermal conductivity of the oxides WO$_2$, WO$_{2.72}$, WO$_{2.90}$ and WO$_3$: (**a**) Total thermal conductivity κ_{tot}; (**b**) Lattice contribution $\kappa_{lat} = \kappa_{tot} - \kappa_{el}$ with κ_{el} from the Wiedemann-Franz law. Mind the different scales of the broken axis.

2.3. Compositions WO$_x$

The phase formation in WO$_x$ samples with varying x is established from the PXRD patterns after the SPS-1420 K/10 min treatment (Figure 4). Semi-logarithmic plotting emphasizes reflections with low intensity. In Table 2 the observed phases are listed.

Table 2. Phases observed from PXRD in WO$_x$ samples prepared with SPS-1420 K/10 min (●—observed, □—not observed).

x	Phases					
	WO$_2$	W$_{18}$O$_{49}$	"WO$_{2.82}$" [a]	W$_{20}$O$_{58}$	W$_{25}$O$_{73}$ [b]	γ/δ-WO$_3$
2	●	□	□	□	□	□
2.50	●	●	□	□	□	□
2.72–2.80	□	●	●	□	□	□
2.82–2.88	□	●	●	●	□	□
2.90	□	□	□	●	□	□
2.92–3	□	□	□	●	●	●

[a] Presumed composition of the new phase. Synchrotron data suggests this to be W$_{12}$O$_{34}$. [b] Phase hard to detect due to strong reflection overlapping.

For $x = 2.50$ (Figure 4) only WO_2 (green squares) and $W_{18}O_{49}$ ($x = 2.72$, brown rhombs) phases are observed as predicted from the phase diagram (Figure 1a).

In the range $2.72 \leq x \leq 2.90$, only $W_{18}O_{49}$ ($x = 2.72$) and $W_{20}O_{58}$ ($x = 2.90$) are expected according to the phase diagram. For $2.72 \leq x \leq 2.80$, $W_{18}O_{49}$ and an additional phase (black arrows) are observed. With increasing O/W-ratio ($2.82 \leq x \leq 2.88$) the expected reflections of $W_{20}O_{58}$ ($x = 2.90$, red dots) appear. The unindexed reflections are most intense for $x = 2.82$. From diffraction data of this sample containing only $W_{18}O_{49}$ and the additional phase, we find many reflection positions fitting to the pentagonal column phase $W_{12}O_{34}$ ($x = 2.83$) [34]. In our case, the parameters of the monoclinic unit cell are slightly changed to $a = 17.229(1)$ Å, $b = 3.782(1)$ Å, $c = 19.496(1)$ Å and $\beta = 105.77(1)$ (Table 1). This is just a rough suggestion; a structure refinement was not successful yet. Distinct reflections of this phase are found also after a comparative synthesis in evacuated silica tubes, which rules out the SPS as cause for its formation (Figure S3). Attempts on synthesis of this phase as single-phase bulk material and further investigations on the structure are pending.

For $2.92 \leq x \leq 2.98$, the PXRD patterns resemble that of pure WO_3 processed in the SPS: reflections of γ-WO_3, δ-WO_3, $W_{20}O_{58}$ ($x = 2.90$) and possibly $W_{25}O_{73}$ ($x = 2.92$) occur. A bluish-violet color of all samples with $2.92 \leq x \leq 2.98$ supports the existence of the latter since $W_{20}O_{58}$ and $W_{25}O_{73}$ are colored deeply blue and violet respectively [14,21]. However, samples with the nominal composition ($x = 2.92$) do not yield single-phase materials. Under SPS conditions, $W_{25}O_{73}$ might not be stable.

2.3.1. Microstructure

From polarized light microscopy (PLM), the grain size of all synthesized WO_x samples is found to be of the order 10–30 μm (Figure 7). A color variation from yellow to red to blue with increasing O/W-ratio x is caused by the different optical reflectivity of the occurring phases. According to SEM images with backscattered electron (BSE) contrast, the samples are homogeneous. Distinct porosity is found only for $x = 2$ and 2.98. Energy dispersive X-ray spectroscopy (EDX) reveals tungsten and oxygen only, but differences of the oxygen content are below the detection limit.

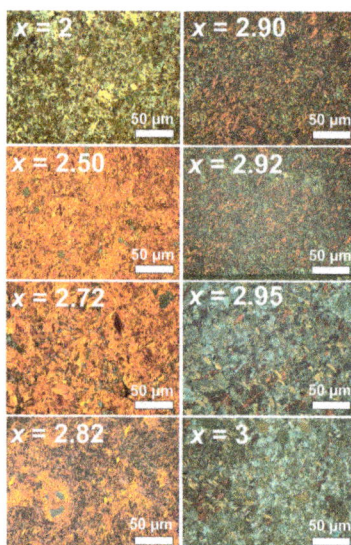

Figure 7. Polarized light microscopy (PLM) of WO_x materials. A yellowish-to-red-to-bluish color variation with increasing O/W-ratio x is observed due to the different optical reflectivity of the occurring phases.

2.3.2. Thermoelectric Properties

The electrical conductivity of the WO$_x$ samples strongly correlates with the O/W-ratio x. When the oxygen concentration is increased to $x \geq 2.90$, there is a continuous decrease of $\sigma(T)$ with minor temperature dependence similar to that of WO$_{2.90}$ (Figure 8a). According to PXRD (Figure 4), next to W$_{20}$O$_{58}$ the nearly insulating phases δ- and γ-WO$_3$ occur in this composition range. Thus, the charge carrier concentration and $\sigma(T)$ mainly depend on the W$_{20}$O$_{58}$ content, which decreases with increasing x. The influence of one further phase, possibly W$_{20}$O$_{73}$ ($x = 2.92$), is not known as of yet.

For decreasing O/W-ratio $x < 2.90$, a metallic behaviour is promoted continuously (Figure 8a). The $\sigma(T)$ approaches that of $x = 2.72$ with the increasing W$_{18}$O$_{49}$ ($x = 2.72$) phase content (Figure 4). No discontinuity of $\sigma(T)$ is found for $x = 2.82$ where the highest amount of "WO$_{2.82}$" is observed. Thus, this phase is electrically conducting. Altogether, the composition $x = 2.90$ is the limit between metallic and nonmetallic behaviour (Figure 8b).

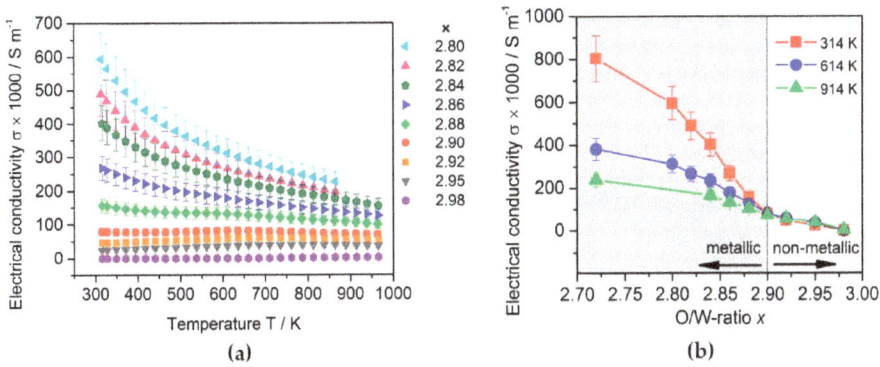

Figure 8. (**a**) Electrical conductivity $\sigma(T)$ of the WO$_x$ samples ($2.80 \leq x \leq 2.98$); (**b**) The temperature dependence of $\sigma(T)$ clearly changes from nonmetallic to metallic at $x \approx 2.90$.

A similar trend is observed for the Seebeck coefficient (Figure 9a). Strong non-monotonic behaviour for the composition $x = 2.98$ results from the large amounts of WO$_3$ in the material (Figure 4) and its therefore near-insulating character during the TE property measurements. The specific influence of the "WO$_{2.82}$"and W$_{25}$O$_{73}$ ($x = 2.92$) phases cannot be considered yet due the lack of the TE properties of the phase-pure material.

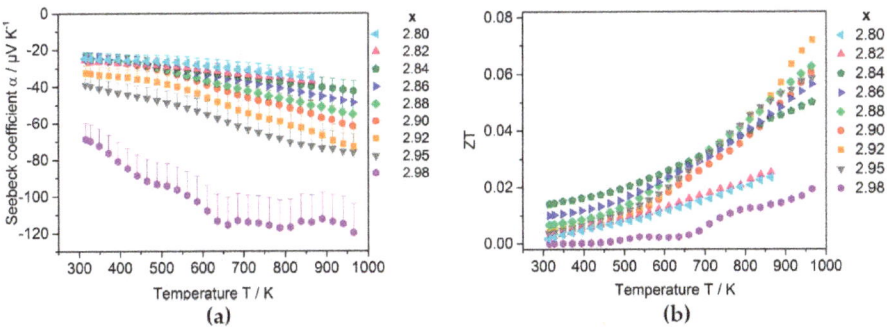

Figure 9. Thermoelectric properties of the WO$_x$ samples ($2.80 \leq x \leq 2.98$): (**a**) Seebeck coefficient $\alpha(T)$ and (**b**) resulting figures of merit ZT. Error of $\alpha(T)$ is asymmetric.

The continuous property change of WO_x materials for $2.80 \leq x \leq 3$ also appears in the thermal conductivity κ_{tot} (Figure 10a). With respect to the uncertainty of the Wiedemann–Franz approximation of κ_{el} (Equations (4) and (5)), κ_{lat} is very similar for all samples $x < 2.98$ (Figure 10b). In the composition range $2.90 \leq x \leq 2.95$ where CS plane phases are found (Table 2), the lattice contribution κ_{lat} appears almost temperature-independent. For $T > 550$ K these samples show no significant reduction of κ_{lat} regarding the WO_3 reference sample with respect to the occuring errors. The exceptionally low value for $x = 2.98$ might be the result of porosity also found by microstructure analysis. However, in the three-phase region ($2.82 \leq x \leq 2.88$) of the system (Table 2), a significantly reduced κ_{lat} is found for $x = 2.84$ up to $T \leq 700$ K. This tendency is also noticable from a plot of κ_{lat} vs. the O/W-ratio x (Figure 10b inset). It indicates that at high temperatures, the lowest κ_{lat} is achieved for WO_x materials with multiple phases due to enhanced phonon scattering at the interfaces.

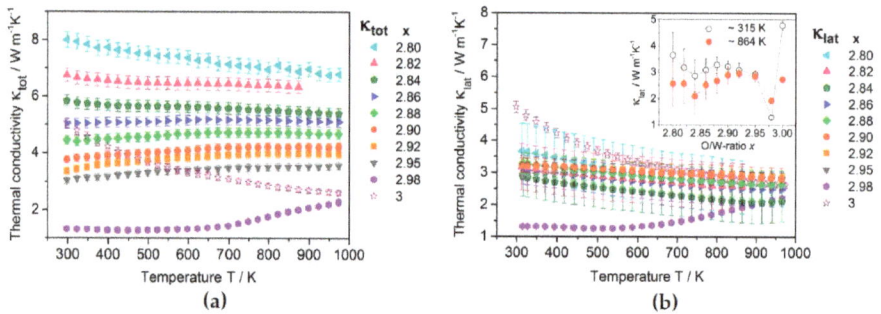

Figure 10. Thermal conductivity of the WO_x samples ($2.80 \leq x \leq 3$): (**a**) Total thermal conductivity $\kappa_{tot}(T)$; (**b**) Lattice contribution $\kappa_{lat} = \kappa_{tot} - \kappa_{el}$ with κ_{el} from the Wiedemann-Franz law. The inset shows the dependency $\kappa_{lat}(x)$ for 315 K and 864 K.

The resulting ZT for the WO_x compositions are very similar to that of $WO_{2.90}$ (Figure 9b). For $x = 2.92$, the highest $ZT = 0.072$ is reached at 963 K. An exceptional low ZT is reached for $x = 2.82$ and $x = 2.98$ due to the high κ_{tot} and low $\sigma(T)$ respectively. However, none of the compositions shows a ZT maximum in the measured temperature range as of yet.

3. Materials and Methods

3.1. Materials

The purchased powders of starting materials WO_3 (Alfa Aesar, 99.998 wt %, 10–20 μm), WO_2 (Sigma-Aldrich, 99.99 wt %, <150 μm) and elemental W (Chempur, 99.9 wt %, 8–9 μm) were analyzed regarding crystalline contaminations with PXRD. For the preparation of WO_x samples ($2.50 \leq x \leq 2.98$), WO_3 was manually mixed for 15 min under argon atmosphere with appropriate amounts of W.

3.2. Spark Plasma Sintering

For spark plasma sintering (SPS), graphite dies with diameters of 8 mm or 10 mm (Figure 2b) and a graphite lining were filled with ca. 1 g of starting mixture under argon atmosphere, and processed with a SPS-515 ET Sinter Lab apparatus (Fuji Electronic Industrial Co. Ltd., Tsurugashima, Japan).

3.3. Powder X-ray Diffraction

The starting materials and SPS-processed samples were examined with powder X-ray diffraction (PXRD) on the Guinier camera G670 (HUBER Diffraktionstechnik GmbH & Co., KG, Rimsting, Germany) with Cu-$K_{\alpha 1}$ radiation ($\lambda = 1.540598$ Å, graphite monochromator, $5° \leq 2\theta \leq 100°$, $\Delta 2\theta = 0.005°$). All PXRD data was compared to the theoretical patterns of known phases (Table 1).

A LaB_6 standard was added to single-phase materials for a subsequent cell parameter determination using the least-square method in the WinCSD software package [38].

3.4. High-Resolution Transmission Electron Microscopy

For high-resolution transmission electron microscopy (HR-TEM), a sample was ground to fine powder and dispersed in methanol. The suspension was loaded on a 100-mesh hexagonal copper grid Quantifoil S7/2 (Quantifoil Micro Tools GmbH, Jena, Germany), which was covered beforehand with a carbon film (2 nm). After a complete drying, HR-TEM imaging was performed on a Tecnai F30 (FEI Technologies Inc., Hillsboro, OR, USA) with a field-emission gun at an acceleration voltage of 300 kV. The point resolution amounted to 1.9 Å, and the information limit amounted to ca. 1.2 Å. The microscope was equipped with a wide-angle slow-scan CCD camera (MultiScan, 2 k × 2 k pixels; Gatan Inc., Pleasanton, CA, USA). TEM images were analyzed with the DigitalMicrograph software (Gatan Inc., Pleasanton, CA, USA).

3.5. Thermoelectric Properties

TE properties were measured for the reference samples x = {2, 2.72, 2.90, 3} and the WO_x samples in the range $2.82 \leq x \leq 2.98$. Samples previously powdered for PXRD were compacted with SPS-1320 K/10 min and a heating rate of 75 K·min^{-1}. Subsequently, the mass density ρ was determined with Archimedean method (Table S1).

The thermal diffusivity $D(T)$ was measured by laser flash analysis (LFA) on a LFA 457 MicroFlash (NETZSCH GmbH & Co. Holding KG, Selb, Germany) in vacuum between room temperature and 963 K in steps of 25 K. Subsequent polishing removed a subtle yellow stain from the pellet surface which occurred during the LFA. Differential scanning calorimetry (DSC) measurements of the specific heat capacity $c_p(T)$ yielded values that for all WO_x samples were identical to WO_3 with respect to the measurement error (Figure S4). Thus, for the calculation of the total thermal conductivity

$$\kappa_{tot}(T) = \rho c_p(T) D(T) \tag{3}$$

(T—absolute temperature) of WO_x and WO_2 there were used theoretic $c_p(T)$ curves for WO_3 and WO_2, respectively [39,40]. After an Archimedian density determination, the microstructure was analyzed with polarized light microscopy (LM), scanning electron microscopy (SEM) and energy dispersive X-ray spectroscopy (EDX). Tetragonal bars with the dimensions $a \approx 1.5$ mm and $b \approx 6$–8 mm were cut from the pellets using a wire saw (Figure 2c). The measurements of the electrical conductivity $\sigma(T)$ and Seebeck coefficient $\alpha(T)$ were performed perpendicular to the SPS pressure direction along the b-edge on a ZEM-3 (ULVAC-RIKO, Munich, Germany) under low pressure helium from RT to 863 K or 963 K in steps of 25 K. The samples were polished before every measurement. The electronic contribution κ_{el} to the total thermal conductivity and consequently the lattice contribution $\kappa_{lat} = \kappa_{tot} - \kappa_{el}$ were calculated with the Wiedemann–Franz equation

$$\kappa_{el}(T) = L(T)\sigma(T)T \tag{4}$$

Since for semiconductors the Lorenz number $L(T)$ can strongly deviate from the constant value $L = 2.4453 \times 10^{-8}$ WΩ·K^{-2} the approximation

$$L(T) = 1.5 + \exp\left[-|\alpha(T)|/116\right] \tag{5}$$

was used, which is based on the single parabolic band model with acoustic phonon scattering [41]. Comparative results obtained with constant L show only minor deviations and all trends remain the same. For a comparison with other TE materials, the figure-of-merit ZT was calculated according to

$$ZT = [\alpha^2(T)\sigma(T)T]/\kappa_{tot}(T). \tag{6}$$

4. Conclusions

According to powder X-ray diffraction (PXRD), a new phase termed as "$WO_{2.82}$" is observed in the composition range $2.72 \leq x \leq 2.90$ in addition to the expected pentagonal column (PC) phase $W_{18}O_{49}$ ($x = 2.72$) and crystallographic shear (CS) plane phase $W_{20}O_{58}$ ($x = 2.90$). The reflection positions of "$WO_{2.82}$" in the synchrotron pattern of a two-phase $W_{18}O_{49}$/"$WO_{2.82}$" sample somewhat fit the PC phase $W_{12}O_{34}$ ($x = 2.83$) with slightly changed lattice parameters. The synthesis as single-phase material and structure refinement still failed, which indicates that the phase is metastable and its structure needs to be further analyzed.

Single-phase $WO_{2.90}$ material is directly obtained from SPS-1420 K/10 min. Both powder X-ray diffraction (PXRD) and high-resolution transmission electron microscopy (HR-TEM) show disorder of the CS planes. Single-phase $WO_{2.70}$ material is obtained from SPS-1420 K/35 min followed by grinding and SPS-1420 K/5 min.

Electronic transport properties of WO_x materials ($2.80 \leq x \leq 2.98$) from the routine SPS-1420 K/10 min reveal a continuous tunability, practically independent from the phases present in the samples. Both the electrical conductivity $\sigma(T)$ and the thermal conductivity $\kappa_{tot}(T)$ indicate an increase of the charge carrier concentration with increasing oxygen deficiency. Here, the composition $WO_{2.90}$ is the limit between metallic ($x < 2.90$) and nonmetallic behaviour ($x \geq 2.90$).

The multi-phase character of these samples found by PXRD is crucial for the thermal conductivity. Significant reduction of κ_{lat} regarding WO_3 is found up to 700 K for the three-phase material $WO_{2.84}$. Thus, for $x < 2.90$ the introduction of multiple phases is a way for reducing the thermal conductivity due to increased phonon scattering at the phase interfaces. The formation of the PC phases in this composition range might have additional influence. In contrast, for $2.90 \leq x \leq 3$, which is the typical range for the formation of CS plane phases, a reduction of κ_{lat} is observed only below 550 K. For high-temperature applications, CS planes might be less appropriate phonon scattering centers.

Supplementary Materials: The following are available online at www.mdpi.com/2073-4352/7/9/271/s1. Figure S1: Electrical conductivity $\sigma(T)$ of the reference materials compositions WO_2, $WO_{2.72}$ and $WO_{2.90}$; Figure S2: Seebeck coefficient $\alpha(T)$ and resulting figures of merit ZT of the reference materials WO_2, $WO_{2.72}$ and $WO_{2.90}$; Figure S3: PXRD patterns of $WO_{2.90}$ obtained from heating in an evacuated silica tube at 1370 K for 72 h and from SPS synthesis at similar temperature for 3 h; Figure S4: Specific heat capacity $c_p(T)$ of WO_x materials; Table S1: Mass density ρ_{theo} of tungsten oxide phases calculated from the molar mass M and unit cell volume V. References [42–44] are cited in the supplementary materials.

Acknowledgments: This research was funded by the Deutsche Forschungsgemeinschaft (DFG) through the priority program SPP1959 "Manipulation of Matter Controlled by Electric and Magnetic Fields: Towards Novel Synthesis and Processing Routes of Inorganic Materials". We acknowledge Yurii Prots for performing the PXRD experiments. Thanks also to Markus Schmidt and Vicky Süß for measurements of the specific heat capacity.

Author Contributions: The project was developed by Igor Veremchuk. Felix Kaiser performed the syntheses, and evaluated the data. Measurements of the transport were performed by Igor Veremchuk. Ulrich Burkhardt analyzed the microstructure. TEM studies were performed by Paul Simon. Important discussions regarding the data interpretation were contributed by Yuri Grin, and Bernd Kieback, who acted as mentor commissioned by the TU Dresden.

Conflicts of Interest: The authors declare no conflicts of interest.

References

1. Terasaki, I.; Sasago, Y.; Uchinokura, K. Large thermoelectric power in $NaCo_2O_4$ single crystals. *Phys. Rev. B* **1997**, *56*, R12685–R12687. [CrossRef]
2. Snyder, G.J.; Toberer, E.S. Complex thermoelectric materials. *Nat. Mater.* **2008**, *7*, 105–114. [CrossRef] [PubMed]
3. Walia, S.; Balendhran, S.; Nili, H.; Zhuiykov, S.; Rosengarten, G. Transition metal oxides—Thermoelectric properties. *Prog. Mater. Sci.* **2013**, *58*, 1443–1489. [CrossRef]
4. Kieslich, G.; Cerretti, G.; Veremchuk, I.; Hermann, R.P.; Panthöfer, M.; Grin, Y.; Tremel, W. A chemists view: Metal oxides with adaptive structures for thermoelectric applications. *Phys. Status Solidi A* **2016**, *213*, 1–16. [CrossRef]

5. Lu, Z.; Zhang, H.; Lei, W.; Sinclair, D.C.; Reaney, I.M. High-Figure-of-Merit Thermoelectric La-Doped A-Site-Deficient SrTiO₃ Ceramics. *Chem. Mater.* **2016**, *28*, 925–935. [CrossRef]

6. Liang, X. Structure and Thermoelectric Properties of Zinc Based Materials. Doctoral Dissertation, Harvard University, Cambridge, MA, USA, 29 August 2013.

7. Mikami, M.; Ozaki, K. Thermoelectric properties of nitrogen-doped TiO_{2-x} compounds. *J. Phys. Conf. Ser.* **2012**, *379*, 012006. [CrossRef]

8. Feng, B.; Martin, H.-P.; Börner, F.-D.; Lippmann, W.; Schreier, M.; Vogel, K.; Lenk, A.; Veremchuk, I.; Dannowski, M.; Richter, C.; et al. Manufacture and Testing of Thermoelectric Modules Consisting of B_xC and TiO_x Elements. *Adv. Eng. Mater.* **2014**, *16*, 1252–1263. [CrossRef]

9. Martin, H.-P.; Pönicke, A.; Kluge, M.; Rost, A.; Conze, S.; Wätzig, K.; Schilm, J.; Michaelis, A. TiO_x-Based Thermoelectric Modules: Manufacturing, Properties, and Operational Behavior. *J. Electron. Mater.* **2016**, *45*, 1570–1575. [CrossRef]

10. Harada, S.; Tanaka, K.; Inui, H. Thermoelectric properties and crystallographic shear structures in titanium oxides of the Magnèli phases. *J. Appl. Phys.* **2010**, *108*, 083703. [CrossRef]

11. Kieslich, G.; Veremchuk, I.; Antonyshyn, I.; Zeier, W.G.; Birkel, C.S.; Weldert, K.; Heinrich, C.P.; Visnow, E.; Panthöfer, M.; Burkhardt, U.; et al. Using crystallographic shear to reduce lattice thermal conductivity: High temperature thermoelectric characterization of the spark plasma sintered Magnèli phases $WO_{2.90}$ and $WO_{2.722}$. *Phys. Chem. Chem. Phys.* **2013**, *15*, 15399. [CrossRef] [PubMed]

12. Kieslich, G.; Tremel, W. Magnèli oxides as promising *n*-type thermoelectrics. *AIMS Mater. Sci.* **2014**, *1*, 184–190. [CrossRef]

13. Migas, D.; Shaposhnikov, V.; Borisenko, V. Tungsten oxides. II. The metallic nature of Magnèli phases. *J. Appl. Phys.* **2010**, *108*, 093714. [CrossRef]

14. Sundberg, M. The crystal and defect structures of $W_{25}O_{73}$, a member of the homologous series W_nO_{3n-2}. *Acta Cryst. B* **1976**, *32*, 2144–2149. [CrossRef]

15. Sundberg, M. Structure and "oxidation behavior" of $W_{24}O_{70}$, a new member of the {103} CS series of tungsten oxides. *Solid State Chem.* **1980**, *35*, 120–127. [CrossRef]

16. Sundberg, M.; Zakharov, N.D.; Zibrov, I.P.; Barabanenkov, Y.A.; Filonenko, V.P.; Werner, P. Two high-pressure tungsten oxide structures of W_3O_8 stoichiometry deduced from high-resolution electron microscopy images. *Acta Cryst. B* **1993**, *49*, 951–958. [CrossRef]

17. Barabanenkov, Y.A.; Zakharov, N.D.; Zibrov, I.P.; Filonenko, V.P.; Werner, P. High-pressure phases in the system W-O-I. Structure of $WO_{1.09}$ by HRTEM. *Acta Cryst. B* **1992**, *48*, 572–577. [CrossRef]

18. McColm, I.J.; Steadman, R.; Wilsoni, S.J. Iron-promoted phases in the tungsten-oxygen system. *J. Solid State Chem.* **1978**, *23*, 33–42. [CrossRef]

19. Stoneham, A.M.; Durham, P.J. The ordering of crystallographic shear planes: Theory of regular arrays. *Phys. Chem. Solids* **1973**, *34*, 2127–2135. [CrossRef]

20. Wriedt, H.A. O-W (Oxygen-Tungsten). In *Binary Alloy Phase Diagrams*, 2nd ed.; Massalski, T.B., Ed.; ASM International: Materials Park, OH, USA, 1990; Volume 3, pp. 2933–2935.

21. Magneli, A. Structure of β-Tungsten Oxide. *Nature* **1950**, *165*, 356–357. [CrossRef]

22. Gebert, E.; Ackermann, R.J. Substoichiometry of Tungsten Trioxide; the Crystal Systems of $WO_{3.00}$, $WO_{2.98}$, and $WO_{2.96}$. *Inorg. Chem.* **1966**, *5*, 136–142. [CrossRef]

23. Berak, J.M.; Sienko, M.J. Effect of oxygen-deficiency on electrical transport properties of tungsten trioxide crystals. *J. Solid State Chem.* **1970**, *2*, 109–133. [CrossRef]

24. Kieslich, G.; Birkel, C.; Douglas, J.E.; Gaultois, M.; Veremchuk, I.; Seshadri, R.; Stucky, G.D.; Grin, Y.; Tremel, W. SPS-assisted preparation of the Magnèli phase $WO_{2.90}$ for thermoelectric applications. *J. Mater. Chem. A* **2013**, *1*, 13050–13054. [CrossRef]

25. Kieslich, G.; Burkhardt, U.; Birkel, C.; Veremchuk, I.; Douglas, J.E.; Gaultois, M.; Lieberwirth, I.; Seshadri, R.; Stucky, G.D.; Grin, Y.; et al. Enhanced thermoelectric properties of the *n*-type Magnèli phase $WO_{2.90}$: Reduced thermal conductivity through microstructure engineering. *J. Mater. Chem. A* **2014**, *2*, 13492–13497. [CrossRef]

26. Venables, D.; Brown, M. Reduction of tungsten oxides with carbon. Part 1: Thermal analyses. *Thermochim. Acta* **1996**, *282*, 251–264. [CrossRef]

27. Woodward, P.M.; Sleight, A.W.; Vogt, T. Ferroelectric Tungsten Trioxide. *J. Solid State Chem.* **1997**, *131*, 9–17. [CrossRef]

28. Gulbransen, E.A.; Andrew, K.F. Kinetics of the Oxidation of Pure Tungsten from 500 to 1300 C. *J. Electrochem. Soc.* **1960**, *107*, 619–628. [CrossRef]

29. Bamwenda, G.R.; Arakawa, H. The Visible Light Induced Photocatalytic Activity of Tungsten Trioxide Powders. *Appl. Catal. A Gen.* **2001**, *210*, 181–191. [CrossRef]

30. Bolzan, H.; Kennedy, B.; Howard, C. Neutron Powder Diffraction Study of Molybdenum and Tungsten Dioxides. *Aust. J. Chem.* **1995**, *48*, 1473–1477. [CrossRef]

31. Magneli, A. Crystal structure studies on β–tungsten oxide. *Ark. Kemi* **1949**, *1*, 223–230.

32. Viswanathan, K.; Brandt, K.; Salje, E. Crystal structure and charge carrier concentration of $W_{18}O_{49}$. *J. Solid State Chem.* **1981**, *36*, 45–51. [CrossRef]

33. Lamire, M.; Labbe, P.; Goreaud, M.; Raveau, B. Refining and new analysis of $W_{18}O_{49}$ structure. *Rev. Chim. Miner.* **1987**, *24*, 369–381.

34. Sundberg, M. Structure determination from HREM images: Application to a new binary tungsten oxide. *Chem. Scr.* **1979**, *14*, 161–166.

35. Magnéli, A. Structures of the ReO_3-type with recurrent dislocations of atoms: 'homologous series' of molybdenum and tungsten oxides. *Acta Cryst.* **1953**, *6*, 495–500. [CrossRef]

36. Vogt, T.; Woodward, P.M.; Hunter, B.A. The High-Temperature Phases of WO_3. *J. Solid State Chem.* **1999**, *144*, 209–215. [CrossRef]

37. Molenda, J.; Kubik, A. Transport properties and reactivity of tungsten trioxide. *Solid State Ion.* **1999**, *117*, 57–64. [CrossRef]

38. Akselrud, L.; Grin, Y. WinCSD: Software package for crystallographic calculations (Version 4). *J. Appl. Cryst.* **2014**, *47*, 803–805. [CrossRef]

39. NIST Chemical Kinetics Database: Tungsten Oxide (WO_3). Available online: http://kinetics.nist.gov/janaf/html/O-065.html (accessed on 20 April 2017).

40. NIST Chemical Kinetics Database: Tungsten Oxide (WO_3). Available online: http://kinetics.nist.gov/janaf/html/O-047.html (accessed on 20 April 2017).

41. Hyun-Sik, K.; Gibbs, Z.M.; Tang, Y.; Wang, H.; Snyder, G.J. Characterization of Lorenz number with Seebeck coefficient measurement. *APL Mater.* **2015**, *3*, 041506. [CrossRef]

42. Wadsley, H.D. Nonstoichiometric Metal Oxides—Order and Disorder. In *Nonstoichiometric Compounds*; Gould, R.D., Ed.; American Chemical Society: Washington, DC, USA, 1963; pp. 23–36. [CrossRef]

43. Salje, E.K.; Rehmann, S.; Pobell, F.; Morris, D.; Knight, K.S.; Herrmannsdörfer, T.; Dovey, M.T. Crystal structure and paramagnetic behaviour of ε-WO_{3-x}. *J. Phys. Condens. Matter* **1997**, *9*, 6563–6577. [CrossRef]

44. Woodward, P.M.; Sleight, A.W.; Vogt, T. Structure refinement of triclinic tungsten trioxide. *J. Phys. Chem. Solids* **1995**, *56*, 1305–1315. [CrossRef]

© 2017 by the authors. Licensee MDPI, Basel, Switzerland. This article is an open access article distributed under the terms and conditions of the Creative Commons Attribution (CC BY) license (http://creativecommons.org/licenses/by/4.0/).

crystals

MDPI

Article

Enhanced Thermoelectric Performance of Te-Doped Bi$_2$Se$_{3-x}$Te$_x$ Bulks by Self-Propagating High-Temperature Synthesis

Rui Liu, Xing Tan, Guangkun Ren, Yaochun Liu, Zhifang Zhou, Chan Liu, Yuanhua Lin * and Cewen Nan

State Key Laboratory of New Ceramics and Fine Processing, Tsinghua University, Beijing 100084, China; liur15@mails.tsinghua.edu.cn (R.L.); tanx14@mails.tsinghua.edu.cn (X.T.); rgk13@mails.tsinghua.edu.cn (G.R.); liuyaoch@126.com (Y.L.); zhifangzhou@163.com (Z.Z.); liuchan16@mails.tsinghua.edu.cn (C.L.); cwnan@mail.tsinghua.edu.cn (C.N.)

* Correspondence: linyh@mail.tsinghua.edu.cn; Tel.: +86-10-6277-3741

Academic Editor: George S. Nolas
Received: 12 June 2017; Accepted: 21 August 2017; Published: 28 August 2017

Abstract: Polycrystalline Bi$_2$Se$_{3-x}$Te$_x$ (x = 0~1.5) samples were prepared by self-propagating high-temperature synthesis (SHS) combined with spark plasma sintering (SPS) and their thermoelectric properties were investigated. The SHS-SPS process can shorten the time with few energy consumptions, and obtain almost pure Bi$_2$Se$_3$-based phases. Consequently, the Se vacancies and anti-site defects contribute to the converged carrier concentration of ~2 × 10^{19} cm^{-3} while the increased carrier effective mass enhances the Seebeck coefficient to more than −158 μV K^{-1} over the entire temperature range. The lattice thermal conductivity is suppressed from 1.07 Wm^{-1} K^{-1} for the pristine specimen to ~0.6 Wm^{-1} K^{-1} for Te-substitution samples at 300 K because of point defects caused by the difference of mass and size between Te and Se atoms. Coupled with the enhanced power factor and reduced lattice thermal conductivity, a high ZT of 0.67 can be obtained at 473 K for the Bi$_2$Se$_{1.5}$Te$_{1.5}$ sample. Our results reveal that Te-substitution based on the SHS-SPS method is highly-efficient and can improve the thermoelectric properties of Bi$_2$Se$_3$-based materials largely.

Keywords: Bi$_2$Se$_{3-x}$Te$_x$; thermoelectric; SHS; solid solution

1. Introduction

With increasing attention on the environmental protection and renewable resources, thermoelectric (TE) instruments, which can directly convert heat into electricity, are considered as a potential solution for harness waste heat [1–3]. Considerable numbers of efforts have been devoted to improving the energy conversion efficiency and the stability of the TE materials [4]. The conversion efficiency depends positively on the dimensionless figure of merit, $ZT = S^2\sigma T/\kappa$, where S is the Seebeck coefficient, σ is the electrical conductivity, κ is the thermal conductivity and T is the absolute temperature, respectively [5]. To maximize the ZT value of a kind of material, a large Seebeck coefficient, electrical conductivity and low thermal conductivity are needed. However, these parameters have a strong coupling with each other, which makes it a challenging task to enhance ZT significantly. Motivated by achieving high thermoelectric performance, multiple methods have been adopted [6]. Many studies so far have focused on the atomic or molecular scales such as doping or alloying to enhance carrier concentration or carrier mobility and thus electrical conductivity to optimize TE performance [7–10]. To enhance Seebeck coefficient while maintaining high electrical conductivity, manipulating the band structure offers a new guideline [11]. Meanwhile, effective alteration at nanometer or mesoscopic scales including the quantum confinement [12] and energy filtering effect [13] can drastically elevate electrical properties.

And developing multi-scale microstructures can obtain a lower thermal conductivity, which is caused by the phonons scattering from high to low frequencies [14]. Besides, further efforts have been made to explore new TE materials and new synthesis methods [15–17].

Bismuth selenide (Bi_2Se_3) is a V-VI semiconductor with a narrow band gap of ~0.3 eV. Several excellent work describing its application of optical recording system [18] and photoelectrochemical devices [19] can be found elsewhere. Because of good TE properties in the mid-temperature, bismuth chalcogenides gained more attention [20–23] in thermoelectrics. Bi_2Se_3 has a rhombohedral layered structure, where Se-Se layers are bonded by van der Waals [24]. On the basis of the weak inter-layer bonding, Sun et al. have reported on the enhancement of the thermoelectric properties of Bi_2Se_3 by the interlayer Cu doping [22]. So far, Bi-Te-Se crystals could be fabricated through the zone melting method [25]. Bi_2Se_3 nanostructures have been synthesized by solvothermal method and ZT of 0.096 was obtained at 523 K [26]. Similarly, for $Bi_2Se_{3-x}Te_x$ ($x \leq 1.5$), Liu prepared by ball milling, but only achieved ZT of ~0.3 [9]. Nonetheless, the mentioned methods of zone melting, solvothermal method and ball milling et al. are time and energy consuming. Meanwhile, the ZT of Bi_2Se_3-based materials is not large enough to meet the requirements of application in mass production. In contrast, self-propagating high-temperature synthesis (SHS) has been proved to be an efficient method to prepare the TE materials alternatively. When the heating rate and temperature are high enough, the reaction wave appears. The heat generated by the exothermic reaction can maintain the whole combustion process, which is exceptionally fast. It shortens the time with few consumptions and can be easily adopted in the commercial application [27]. A wide range of TE materials have been synthesized successfully by this method, including Cu_2Se, $BiCuSeO$, Cu_3SbSe_3 and so on [27–30].

Previous work showed that Bi_2Se_3 and Bi_2Te_3 can be prepared by SHS method [31]. However, there are few studies focusing on the thermoelectric properties of $Bi_2Se_{3-x}Te_x$ ($x \leq 1.5$) prepared by combustion method. In this work, we successfully synthesized $Bi_2Se_{3-x}Te_x$ ($x = 0$, 0.3, 0.6, 0.9, 1.2, 1.5) via the SHS method followed by spark plasma sintering (SPS) and studied the thermoelectric properties from 300 K to 593 K. Our results show that the highest power factor ($PF = S^2\sigma$) can achieve 11.2 $\mu Wcm^{-1} K^{-2}$ for $Bi_2Se_{1.5}Te_{1.5}$ at 300 K and the lattice thermal conductivity (κ_L) could be reduced to the lowest value of 0.35 $Wm^{-1} K^{-1}$ at 593 K via Te alloying for $Bi_2Se_{2.1}Te_{0.9}$. The ZT of ~0.67 is finally achieved at 473 K for $Bi_2Se_{1.5}Te_{1.5}$, demonstrating the potential application for energy conversion in the mid-temperature. And SHS process will have more hopeful prospects in commercial applications.

2. Experimental Procedures

In the initial stage, Bi (99.99%, Aladdin), Se (99.99%, Aladdin), and Te (99.99%, Aladdin) powders were mixed meticulously in stoichiometric amounts. Then the mixture was cold-pressed into pellets with the diameter of 20 mm. The SHS process was started by heating the bottom of the pellets with a hand torch in the air. Once ignited, the hand torch is removed immediately. The heat generated by the combination reaction kept the combustion process propagating until it was finished in several seconds. Then the pellets were grounded into fine powders carefully by hand. The powders were then sintered into pellets of ϕ 12.7 mm by SPS (Sumitomo Coal Mining Co., Ltd., Tokyo, Japan) at the temperature of 593 K for 5 min under a uniaxial pressure of 40 MPa.

The phase structures were investigated by X-ray diffraction (XRD, RINT2000, Rigaku, Tokyo, Japan) analysis. The morphology and composition of cross-sectional bulks were checked by field-emission scanning electron microscopy (FESEM) (LEO1530, Oxford Instruments, Oxford, UK). The electrical properties including electrical conductivity and Seebeck coefficient were measured from room temperature to 593 K by ZEM-3 (ULVAC, Kanagawa, Japan). The van der Pauw method was used in an Eastchanging Hall measurement station to measure Hall coefficient (R_H). The carrier concentration (n) and mobility (μ) were estimated by the equation n = $1/eR_H$ and $\mu = \sigma R_H$. To ensure the accuracy, the samples were polished to be thinner than 0.5 mm for the measurements. The total thermal conductivity is determined by the equation $\kappa = DC_p\rho$, where D is thermal diffusivity, C_p is specific heat and ρ is the density of the bulks. The thermal diffusivity was obtained by the laser flash method and the specific heat was calculated by the Dulong-Petit relation. The density of the bulks was derived with Archimedes method.

3. Results and Discussion

Figure 1a is the XRD result of all the $Bi_2Se_{3-x}Te_x$ samples with x = 0~1.5. All the major peaks in the XRD patterns correspond to a standard card, Bi_2Se_3, PDF #33-0214. The additional small peaks can be identified as Bi_2O_2Se (PDF #29-0237), which is possibly generated by oxidation during the ultra-fast combustion process in the air. In this work, we assume that all the samples contain the same amount of Bi_2O_2Se, and we neglect the effect of existence of Bi_2O_2Se due to its small amount (small peaks in the XRD result). In Figure 1b, the lattice parameters were calculated according to the position of XRD peaks. With increasing Te content, the lattice constants a and c increase linearly, which indicates Te can successfully substitute for Se atoms to form solid solution by SHS process in a short time. Figure 1c–e show the morphology of cross-sectional bulks (x = 0, 0.3, 1.5). All the samples were sintered well with high density (94% or above). The layer structure can be seen clearly in the $Bi_2Se_{3-x}Te_x$ bulks.

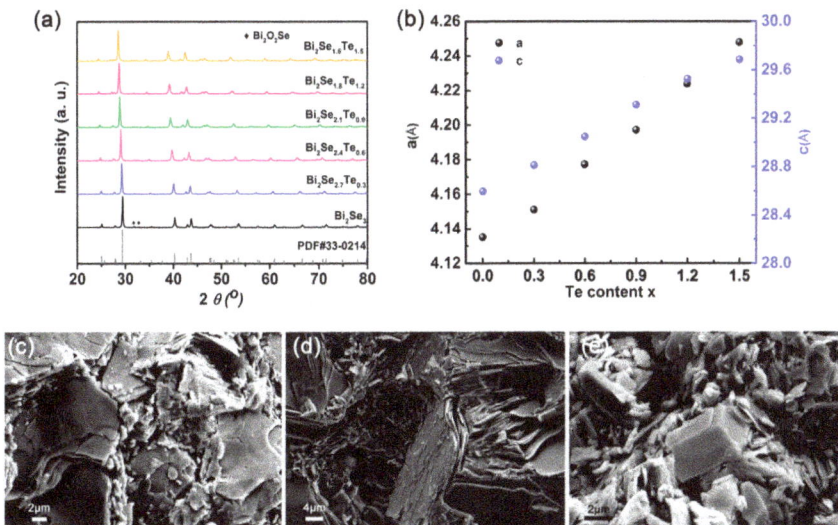

Figure 1. (a) XRD patterns and (b) lattice parameters of sintered $Bi_2Se_{3-x}Te_x$ bulk samples; field-emission scanning electron micrographs of $Bi_2Se_{3-x}Te_x$, for which, (c) x = 0; (d) x = 0.3; and (e) x = 1.5.

Figure 2 shows the temperature dependence of electrical conductivity and Seebeck coefficient. The electrical conductivity σ (in Figure 2a) of pristine Bi_2Se_3 maintains at about 400 Scm^{-1} from 300 K to 593 K, which is much higher than Bi_2Se_3 prepared by other method [22]. Se is much easier to evaporate during the combustion process because of low energy of evaporation and thus it leaves Se vacancies and free electrons, which may contribute to higher electrical conductivity. This can be indicated in the following equation:

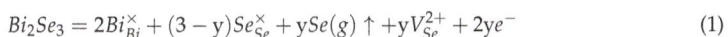

$$Bi_2Se_3 = 2Bi_{Bi}^{\times} + (3-y)Se_{Se}^{\times} + ySe(g) \uparrow + yV_{Se}^{2+} + 2ye^{-} \tag{1}$$

As the Te content increases (x > 0), the electrical conductivity at 300 K initially increases to ~870 Scm^{-1} because of increased carrier concentration (Table 1), then decreases to ~400 Scm^{-1} owing to the change of carrier mobility, which is much lower than the pristine Bi_2Se_3 (Table 1). The carrier concentration increases may be a result of increasing anti-site defects (Bi_{Te}^-) [31], which is caused by the fact that Bi can jump from Bi-site to Te-site easily because of small difference in electronegativity [10], as indicated in Equation (2).

$$Bi_2Te_3 = (2 - \frac{2}{5}z)Bi_{Bi}^{\times} + (3 - z)Te_{Te}^{\times} + zTe(g) \uparrow + (\frac{2}{5}zV_{Bi}^{3-} + \frac{3}{5}zV_{Te}^{2+}) + \frac{2}{5}zBi_{Te}^- + \frac{2}{5}zh^+ \qquad (2)$$

On the contrary, the number of Se vacancies V_{Se}^{2+} will be fewer due to the increasing Te content. The decreased Se vacancies and increased anti-site defects make the carrier concentration converges to about 2×10^{19} cm^{-3}. The carrier mobility decreases with higher Te content at 300 K in general due to the enhanced alloy scattering. Interestingly, we found μ of Bi$_2$Se$_{1.5}$Te$_{1.5}$ (x = 1.5) is slightly larger than Bi$_2$Se$_{1.8}$Te$_{1.2}$ (x = 1.2) at the room temperature, which is possibly caused by the intrinsic high mobility of Bi$_2$Te$_3$ [31].

Table 1. Actual composition, carrier concentration (n), carrier mobility (μ), carrier effective mass (m^*), Seebeck coefficient (S), Lorenz constant (L), lattice thermal conductivity (κ_L), κ_L/κ, and density of Bi$_2$Se$_{3-x}$Te$_x$ samples at 300 K.

x	n (10^{18} cm^{-3})	μ (cm^2 V^{-1} s^{-1})	m^*/m_0	S (μV K^{-1})	L (10^{-8} V^2 K^{-2})	κ_L (Wm^{-1} K^{-1})	κ_L/κ	Density (g cm^{-3})
0.0	5.94	444.48	0.19	−118.23	1.83	1.07	81.7%	7.01
0.3	17.31	309.19	0.25	−73.65	2.06	0.80	59.4%	6.92
0.6	21.94	209.36	0.31	−79.95	2.03	0.62	57.4%	6.77
0.9	24.42	149.33	0.45	−107.68	1.88	0.52	60.4%	6.65
1.2	24.37	97.45	0.53	−126.49	1.81	0.60	74.0%	6.83
1.5	20.73	134.96	0.60	−158.72	1.71	0.59	71.6%	6.96

As shown in Figure 2b, the negative Seebeck coefficient of all the samples indicates the dominance of electrons in the transport process. Generally, the value of Seebeck coefficient can be estimated by the equation [32]:

$$|S| = \frac{8\pi^2 k_B^2 T}{3eh^2} m_d^* (\frac{\pi}{3n})^{2/3} \qquad (3)$$

where e, k_B, T, h, m_d^*, and n are the carrier charge, Boltzmann constant, absolute temperature, Planck constant, the effective mass of the carrier, and carrier concentration. As shown in the formula, because of largely enhanced carrier concentration with increasing Te content (x < 0.9) at 300 K, the Seebeck coefficient decreases. Then the Seebeck coefficient was improved due to the larger carrier effective mass at 300 K (Table 1). It should be noticed that each sample with x $\geq$ 0.9, as the temperature increases, the value of Seebeck coefficient first increases then decreases, which is caused by the intrinsic excitations. The highest Seebeck coefficient of −180 μV K^{-1} is achieved at 473 K for the Bi$_2$Se$_{1.5}$Te$_{1.5}$ sample. The Bi$_2$Se$_{1.5}$Te$_{1.5}$ sample attains the largest effective mass of ~0.60 m$_0$, which is in accord with the difference of Seebeck coefficient with different Te contents at room temperature.

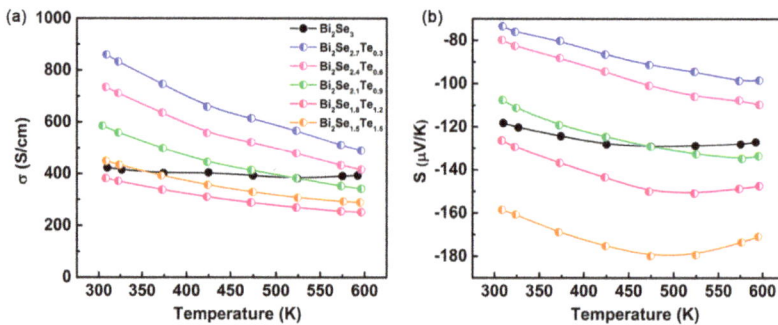

Figure 2. Temperature dependence of (**a**) electrical conductivity and (**b**) Seebeck coefficient for Bi$_2$Se$_{3-x}$Te$_x$ samples.

The variation of power factor ($PF = S^2\sigma$) with increasing temperature of all the samples is shown in Figure 3. The $Bi_2Se_{1.5}Te_{1.5}$ sample reaches the highest PF of 11.2 $\mu Wcm^{-1} K^{-2}$ at room temperature, which is almost twice higher than that of pristine Bi_2Se_3. But it drops to about 8 $\mu Wcm^{-1} K^{-2}$ at 593 K owing to the decreased electrical conductivity and Seebeck coefficient.

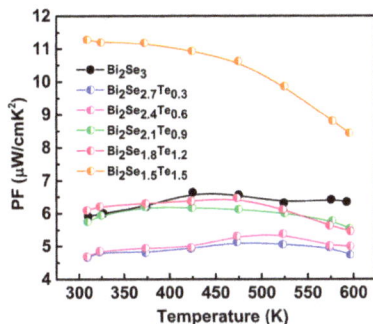

Figure 3. The temperature dependence of power factor for $Bi_2Se_{3-x}Te_x$ samples.

Figure 4a illustrates the total thermal conductivity (κ) as a function of temperature from room temperature to 593 K. The κ of pristine Bi_2Se_3 is in the range of 1.04–1.31 $Wm^{-1} K^{-1}$. As x increases to 1.2, the κ drops into the range of 0.76–0.82 $Wm^{-1} K^{-1}$ from 300 K to 593 K. With further increasing Te content, κ is much larger than the sample of x = 1.2. To have a better understanding of the thermal transport properties, κ is subsequently divided into three parts:

$$\kappa = \kappa_e + \kappa_L + \kappa_B \tag{4}$$

where κ_e is electron thermal conductivity, κ_L is lattice thermal conductivity and κ_B is the bipolar thermal conductivity induced by intrinsic excitaions. κ_e can be estimated by Wiedemann-Franz relation:

$$\kappa_e = L\sigma T \tag{5}$$

where L is the Lorenz constant and σ is electrical conductivity. In the single parabolic band model, L depends on the reduced chemical potential and scattering parameter. It can be estimated by fitting the values of the Seebeck coefficient and the room temperature data has been listed in Table 1. The details can be seen elsewhere [33,34]. As mentioned above (Figure 2b), intrinsic excitaions don't occur until 423 K or above. Therefore, the κ_B can be ignored at low temperature in Figure 4b. Consequently, κ_L and the reciprocal temperature, T^{-1}, follow a linear relationship. As shown in Figure 4c, the lattice thermal conductivity drops substantially after alloying. Note that the κ_L of $Bi_2Se_{2.1}Te_{0.9}$ achieves the lowest value of 0.35 $Wm^{-1} K^{-1}$ at 593 K. The effective suppression of the κ_L of could be attributed to point defects caused by the different mass and size between Te and Se atoms. Similar to the previous literature [9], the κ_L rises slightly when x is above 0.9. This may be ascribed to the relatively high κ_L of Bi_2Te_3 [31], whose effect is larger than the point defects. Figure 4d shows the temperature dependence of κ_B. With low Te content (x $\leq$ 0.6), κ_B is almost zero from 300 K to 593 K, because there are no intrinsic excitaions. Intrinsic excitaions occur and κ_B's contribution to κ becomes larger when x $\geq$ 0.9, which is owing to narrower band gap with increasing Te content [35].

The ZT values for all the $Bi_2Se_{3-x}Te_x$ samples (x = 0, 0.3, 0.6, 0.9, 1.2, 1.5) are presented in Figure 5. The enhanced power factor and the effective suppression of lattice thermal conductivity synergistically contribute to the highest ZT value of 0.67 at 473 K for the sample of $Bi_2Se_{1.5}Te_{1.5}$, which is almost twice higher than the pristine Bi_2Se_3. ZT values of Bi_2Se_3 (ball milling) [22], Bi_2Se_2Te (ball milling) [9],

Bi$_2$Te$_3$ (SHS) [31] from the literature are included for comparison. Our results show that the *ZT* value of Te-substituted Bi$_2$Se$_3$-based materials prepared by SHS is much larger than that by other methods.

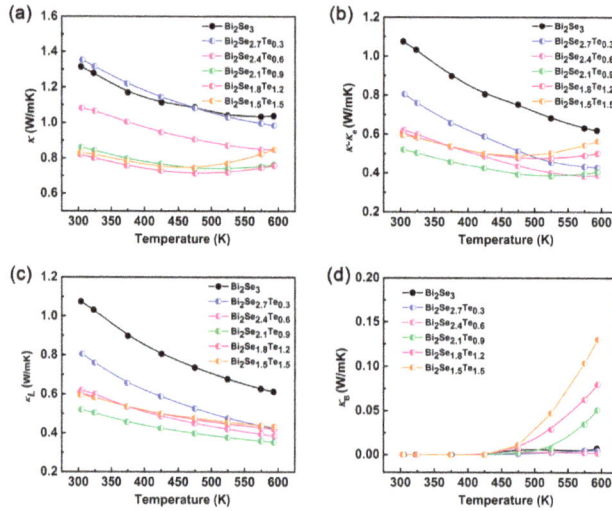

Figure 4. The temperature dependence of the (**a**) total thermal conductivity; (**b**) $\kappa - \kappa_B$; (**c**) lattice thermal conductivity; and (**d**) the bipolar thermal conductivity for Bi$_2$Se$_{3-x}$Te$_x$ samples.

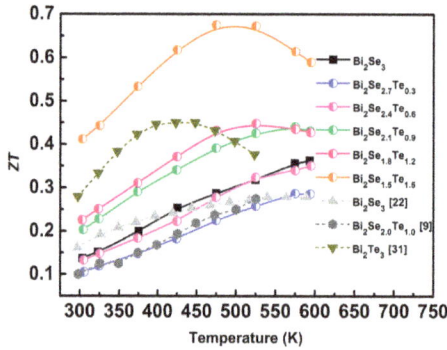

Figure 5. The temperature dependence of ZT for Bi$_2$Se$_{3-x}$Te$_x$ samples.

4. Conclusions

In summary, we have investigated the thermoelectric properties (300–593 K) of Bi$_2$Se$_{3-x}$Te$_x$ samples (x = 0, 0.3, 0.6, 0.9, 1.2, 1.5), which are prepared by SHS-SPS process successfully. Compared with other methods, the SHS-SPS process is much faster and requires less energy, which is desirable in commercial application even though with small amount of second phase Bi$_2$O$_2$Se. Our results show that the power factor of Bi$_2$Se$_{1.5}$Te$_{1.5}$ achieves 11.2 μWcm^{-1} K^{-2} at 300 K by the increased carrier concentration and the enhancement of Seebeck coefficient. The point defects originate from the difference of mass and size between Te and Se atoms significantly suppresses the lattice thermal conductivity. Benefiting from the improved power factor and the decreased lattice thermal conductivity, a high *ZT* of 0.67 can be obtained at 473 K for the sample of Bi$_2$Se$_{1.5}$Te$_{1.5}$, which demonstrates that the Te-substitution via SHS-SPS method is highly-efficient and can enhance the thermoelectric properties of Bi$_2$Se$_3$-based materials.

Acknowledgments: This work was supported by the National Key Research Programme of China, under grant No. 2016YFA0201003, Ministry of Sci & Tech of China through a 973-Project under grant No. 2013CB632506, and National Science Foundation of China under grand No. 51672155 and 51532003.

Author Contributions: Rui Liu performed the experiments and wrote the paper. Guangkun Ren, Yaochu Liu, Cewen Nan and Yuanhua Lin revised the manuscript. Chan Liu, Xing Tan and Zhifang Zhou assisted in experiments.

Conflicts of Interest: The authors declare no conflict of interest.

References

1. Yang, S.H.; Zhu, T.J.; Sun, T.; He, J.; Zhang, S.N.; Zhao, X.B. Nanostructures in high-performance $(GeTe)_x(AgSbTe_2)_{100-x}$ thermoelectric materials. *Nanotechnology* **2008**, *19*, 245707. [CrossRef] [PubMed]
2. Bell, L.E. Cooling, heating, generating power, and recovering waste heat with thermoelectric systems. *Science* **2008**, *321*, 1457–1461. [CrossRef] [PubMed]
3. Tritt, T.M. Holey and unholey semiconductors. *Science* **1999**, *283*, 804. [CrossRef]
4. Rhyee, J.S.; Lee, K.H.; Lee, S.M.; Cho, E.; Kim, S.I.; Lee, E.; Kwon, Y.S.; Shim, J.H.; Kotliar, G. Peierls distortion as a route to high thermoelectric performance in $In_4Se_{3-\delta}$ crystals. *Nature* **2009**, *459*, 965–968. [CrossRef] [PubMed]
5. Nolas, G.S.; Sharp, J.; Goldsmid, J. *Thermoelectrics: Basic Principles and New Materials Developments*; Springer: Berlin, Germany, 2013; Volume 45, p. 111.
6. Snyder, G.J.; Toberer, E.S. Complex thermoelectric materials. *Nat. Mater.* **2008**, *7*, 105–114. [CrossRef] [PubMed]
7. Mehta, R.J.; Zhang, Y.; Karthik, C.; Singh, B.; Siegel, R.W.; Borca-Tasciuc, T.; Ramanath, G. A new class of doped nanobulk high-figure-of-merit thermoelectrics by scalable bottom-up assembly. *Nat. Mater.* **2012**, *11*, 233–240. [CrossRef] [PubMed]
8. Tan, X.; Lan, J.L.; Ren, G.K.; Liu, Y.; Lin, Y.H.; Nan, C.W. Enhanced thermoelectric performance of n-type Bi_2O_2Se by Cl-doping at Se site. *J. Am. Ceram. Soc.* **2017**, *100*, 1494–1501. [CrossRef]
9. Liu, W.S.; Lukas, K.C.; McEnaney, K.; Lee, S.; Zhang, Q.; Opeil, C.P.; Chen, G.; Ren, Z.F. Studies on the Bi_2Te_3-Bi_2Se_3-Bi_2S_3 system for mid-temperature thermoelectric energy conversion. *Energy Environ. Sci.* **2013**, *6*, 552–560. [CrossRef]
10. Liu, W.S.; Zhang, Q.Y.; Lan, Y.C.; Chen, S.; Yan, X.; Zhang, Q.; Wang, H.; Wang, D.Z.; Chen, G.; Ren, Z.F. Thermoelectric property studies on Cu-doped n-type $Cu_xBi_2Te_{2.7}Se_{0.3}$ nanocomposites. *Adv. Energy Mater.* **2011**, *1*, 577–587. [CrossRef]
11. Pei, Y.Z.; Shi, X.Y.; Lalonde, A.; Wang, H.; Chen, L.D.; Snyder, G.J. Convergence of electronic bands for high performance bulk thermoelectrics. *Nature* **2011**, *473*, 66–69. [CrossRef] [PubMed]
12. Hicks, L.D.; Dresselhaus, M.S. Thermoelectric figure of merit of a one-dimensional conductor. *Phys. Rev. B* **1992**, *47*, 16631. [CrossRef]
13. Humphrey, T.E.; Linke, H. Reversible Thermoelectric Nanomaterials. *Phys. Rev. Lett.* **2005**, *94*, 096601. [CrossRef] [PubMed]
14. Biswa, K.; He, J.Q.; Blum, I.D.; Wu, C.I.; Hogan, T.P.; Seidman, D.D.; Dravid, V.P.; Kanatzidis, M.G. High-performance bulk thermoelectrics with all-scale hierarchical architectures. *Nature* **2012**, *489*, 414–418. [CrossRef] [PubMed]
15. Dresselhaus, M.S.; Chen, G.; Tang, M.Y.; Yang, R.; Lee, H.; Wang, D.; Ren, Z.; Fleurial, J.P.; Gogna, P. New directions for low-dimensional thermoelectric materials. *Adv. Mater.* **2007**, *19*, 1043–1053. [CrossRef]
16. Chen, G.; Dresselhaus, M.; Dresselhaus, G.; Fleurial, J.P.; Caillat, T. Recent developments in thermoelectric materials. *Int. Mater. Rev.* **2003**, *48*, 45–66. [CrossRef]
17. Kauzlarich, S.M.; Brown, S.R.; Snyder, G.J. Zintl phases for thermoelectric devices. *Dalton Trans.* **2007**, *21*, 2099–2107. [CrossRef] [PubMed]
18. Watanabe, K.; Sato, N.; Miyaoko, S. New optical recording material for video disc system. *J. Appl. Phys.* **1983**, *54*, 1256. [CrossRef]
19. Waters, J.; Crouch, D.; Raftery, J.; O'Brien, P. Deposition of bismuth chalcogenide thin films using novel single-source precursors by metal-organic chemical vapor deposition. *Chem. Mater.* **2004**, *16*, 3289–3298. [CrossRef]

20. Bayaz, A.A.; Giani, A.; Foucaran, A.; Pascal-Delannoy, F.; Boyer, A. Electrical and thermoelectrical properties of Bi_2Se_3 grown by metal organic chemical vapour deposition technique. *Thin Solid Films* **2003**, *441*, 1–5. [CrossRef]

21. Tang, Z.L.; Hu, L.P.; Zhu, T.J.; Liu, X.H.; Zhao, X.B. High performance n-type bismuth telluride based alloys for mid-temperature power generation. *J. Mater. Chem. C* **2015**, *3*, 10597–10603. [CrossRef]

22. Sun, G.L.; Qin, X.Y.; Li, D.; Zhang, J.; Ren, B.J.; Zou, T.H.; Xin, H.X.; Paschen, S.B.; Yan, X.L. Enhanced thermoelectric performance of n-type Bi_2Se_3 doped with Cu. *J. Alloys Compd.* **2015**, *639*, 9–14. [CrossRef]

23. Kim, D.; Syers, P.; Butch, N.P.; Paglione, J.; Fuhrer, M.S. Ambipolar surface state thermoelectric power of topological insulator Bi_2Se_3. *Nano Lett.* **2014**, *14*, 1701–1706. [CrossRef] [PubMed]

24. Nakajima, S. The crystal structure of $Bi_2Te_{3-x}Se_x$. *J. Phys. Chem. Solids* **1963**, *24*, 479–485. [CrossRef]

25. Jiang, J.; Chen, L.D.; Yao, Q.; Bai, S.Q.; Wang, Q. Effect of TeI4 content on the thermoelectric properties of n-type Bi–Te–Se crystals prepared by zone melting. *Mater. Chem. Phys.* **2005**, *92*, 39–42. [CrossRef]

26. Kadel, K.; Kumari, L.; Li, W.Z.; Huang, Y.Y.; Provencio, P.P. Synthesis and thermoelectric properties of Bi_2Se_3 nanostructures. *Nanoscale Res. Lett.* **2011**, *6*, 57. [CrossRef] [PubMed]

27. Su, X.; Fu, F.; Yan, Y.; Zheng, G.; Liang, T.; Zhang, Q.; Cheng, X.; Yang, D.; Chi, H.; Tang, X.; et al. Self-propagating high-temperature synthesis for compound thermoelectrics and new criterion for combustion processing. *Nat. Commun.* **2014**, *5*, 4908. [CrossRef] [PubMed]

28. Ren, G.K.; Lan, J.L.; Butt, S.; Ventura, K.J.; Lin, Y.H.; Nan, C.W. Enhanced thermoelectric properties in Pb-doped BiCuSeO oxyselenides prepared by ultrafast synthesis. *RSC Adv.* **2015**, *5*, 69878–69885. [CrossRef]

29. Liu, R.; Ren, G.K.; Tan, X.; Lin, Y.L.; Nan, C.W. Enhanced thermoelectric properties of Cu_3SbSe_3-based composites with inclusion phases. *Energies* **2016**, *9*, 816. [CrossRef]

30. Yang, D.W.; Su, X.L.; Yan, Y.G.; Hu, T.Z.; Xie, H.Y.; He, J.; Uher, C.; Kanatzidis, M.G.; Tang, X.F. Manipulating the combustion wave during self-Propagating synthesis for high thermoelectric performance of layered oxychalcogenide $Bi_{1-x}Pb_xCuSeO$. *Chem. Mater.* **2016**, *28*, 4628–4640. [CrossRef]

31. Zheng, G.; Su, X.L.; Liang, T.; Lu, Q.B.; Yan, Y.G.; Uher, C.; Tang, X.F. High thermoelectric performance of mechanically robust n-type $Bi_2Te_{3-x}Se_x$ prepared by combustion synthesis. *J. Mater. Chem. A* **2015**, *3*, 6603–6613. [CrossRef]

32. Koumoto, K.; Funahashi, R.; Guilmeau, E.; Miyazaki, Y.; Weidenkaff, A.; Wang, Y.F.; Wan, C.L. Thermoelectric ceramics for energy harvesting. *J. Am. Ceram. Soc.* **2013**, *96*, 1–23. [CrossRef]

33. Pei, Y.L.; He, J.; Li, J.F.; Li, F.; Liu, Q.J.; Pan, W.; Barreteau, C.; Berardan, D.; Dragoe, N.; Zhao, L.D. High thermoelectric performance of oxyselenides: Intrinsically low thermal conductivity of Ca-doped BiCuSeO. *NPG Asia Mater.* **2013**, *5*, 425–434. [CrossRef]

34. Kumar, G.S.; Prasad, G.; Pohl, R.O. Experimental determinations of the Lorenz number. *J. Mater. Sci.* **1993**, *28*, 4261–4272. [CrossRef]

35. Imamuddin, M.; Dupre, A. Thermoelectric properties of p-type Bi_2Te_3-Sb_2Te_3-Sb_2Se_3 alloys and n-type Bi_2Te_3-Bi_2Se_3 alloys in the temperature range 300 to 600 K. *Phys. Status Solidi A* **1972**, *10*, 415–424. [CrossRef]

© 2017 by the authors. Licensee MDPI, Basel, Switzerland. This article is an open access article distributed under the terms and conditions of the Creative Commons Attribution (CC BY) license (http://creativecommons.org/licenses/by/4.0/).

crystals

MDPI

Article

High Temperature Transport Properties of Yb and In Double-Filled p-Type Skutterudites

Dean Hobbis [1], Yamei Liu [2], Kaya Wei [1], Terry M. Tritt [2] and George S. Nolas [1,*]

[1] Department of Physics, University of South Florida, Tampa, FL 33620, USA; dhobbis@mail.usf.edu (D.H.); kayawei@mail.usf.edu (K.W.)

[2] Department of Physics and Astronomy, Kinard Laboratory, Clemson University, Clemson, SC 29634, USA; yameil@g.clemson.edu (Y.L.); ttritt@clemson.edu (T.M.T.)

* Correspondence: gnolas@usf.edu; Tel.: +1-813-974-2233

Academic Editor: Helmut Cölfen
Received: 18 July 2017; Accepted: 18 August 2017; Published: 23 August 2017

Abstract: Yb and In double-filled and Fe substituted polycrystalline p-type skutterudite antimonides were synthesized by direct reaction of high-purity elements, followed by solid-state annealing and densification by hot pressing. The stoichiometry and filling fraction were determined by both Rietveld refinement of the X-ray diffraction data and energy dispersive spectroscopic analyses. The transport properties were measured between 300 K and 830 K, and basically indicate that the resistivity and Seebeck coefficient both increase with increasing temperature. In both specimens, the thermal conductivity decreased with increasing temperature up to approximately 700 K, where the onset of bipolar conduction was observed. A maximum ZT value of 0.6 at 760 K was obtained for the $Yb_{0.39}In_{0.018}Co_{2.4}Fe_{1.6}Sb_{12}$ specimen.

Keywords: thermoelectric; skutterudite; p-type; figure of merit; double-filled; bipolar diffusion

1. Introduction

Thermoelectric materials research is of current significant interest for improving device performance in order to efficiently convert waste heat into electrical power [1]. Thermoelectric device improvement would result in an expanded array of potential applications, including automobile applications [2]. The efficiency of a thermoelectric material is given by the dimensionless figure of merit $ZT = S^2 T / \rho \kappa$, where S is the Seebeck coefficient, T is the absolute temperature, ρ is the resistivity, and κ is the thermal conductivity. The larger both the average and the peak ZT value are, the better the thermoelectric properties of a material. Both n-type and p-type materials are required in a thermoelectric device; the efficiency of the device is characterized by the combination of both materials' thermoelectric properties.

Skutterudites have been studied extensively—not only due to their encouraging thermoelectric performance at intermediate temperatures, but also due to their good mechanical properties [3–6]. Optimization of p-type skutterudites is difficult due to the relatively small effective mass of holes compared to the effective mass of electrons in these materials [4]; therefore, the optimum carrier concentration of n-type skutterudites is larger than that of p-type, leading to a larger power factor ($S^2\sigma$, where σ is electrical conductivity) [7]. It is well documented that the twelve Sb atoms in the skutterudite crystal structure form relatively large icosahedral cages [1,3,8]; thus, reduction of the lattice thermal conductivity, κ_L, can be achieved by fractional filling of these cages with rare-earth, alkali-earth, or alkali-metal atoms, since these cage-fillers result in the scattering of lattice phonons [3,8]. Yb filling has been shown to be a good filler candidate for skutterudites because of its large mass and small ionic radius, which results in strong phonon scattering. Furthermore, in skutterudites Yb has been shown to have an intermediate valence state (+2~+3), demanding less charge compensation for p-type materials [9]. The thermoelectric properties of n-type (Yb, In) double-filled skutterudite antimonides have been previously reported with a maximum *ZT* of 0.97 [10]. The pursuit of p-type materials is also needed; herein we investigate similar double filling in p-type skutterudites in order to determine their potential for thermoelectric applications.

2. Experimental

The high-purity elements were weighed and loaded into silica crucibles in a N_2 environment inside a glove box to minimize exposure to air. Yb chunks (99.9%, Ames Labs), In foil (99.9975%, Alfa Aesar), Co powder (99.998%, Alfa Aesar), Fe powder (99.998%, Alfa Aesar), and crushed Sb chunks (99.5%, Alfa Aesar) were reacted in the nominal compositions $Yb_{0.4}In_{0.02}Co_3FeSb_{12}$ and $Yb_{0.8}In_{0.02}Co_{2.5}Fe_{1.5}Sb_{12}$ for this study, which were chosen based off of previous studies of Yb single-filled Fe substituted skutterudites [11,12]. The specimens were sealed in a quartz tube under vacuum and reacted in a furnace at 1173 K for 48 h. The tube was removed and allowed to cool to room temperature in air before the specimens were ground into fine powders in a N_2 glove box and cold pressed into pellets. These pellets were again sealed under vacuum in a quartz tube and annealed at 973 K for 7 days. This grinding and annealing process was repeated once more to further encourage homogeneity. The specimens were then finely ground and sieved (325 mesh) before being loaded into a graphite die inside the glove box for hot pressing. The hot pressing conditions for densification were performed under constant N_2 flow at 923 K and 120 MPa for 3 h, resulting in high-density polycrystalline skutterudites as measured by the Archimedes method.

Analyses of the homogeneity and stoichiometry of the specimens were performed by Rietveld refinement of the powder X-ray diffraction (XRD) data using a Bruker D8 Focus Diffractometer in Bragg–Brentano geometry with Cu Kα radiation and a graphite monochromator, and energy dispersive spectroscopy (EDS) using an Oxford INCA X-Sight 7852 equipped scanning electron microscope (SEM, JEOL, JSM-6390LV). The densified pellets were cut with a wire saw for high-temperature transport measurements. A rectangular parallelepiped ($2 \times 2 \times 5$ mm^3) was used for four-probe ρ and S measurements on a ULVAC ZEM-2 system. Thermal diffusivity measurements on a thin disk were performed by the laser flash method on a NETZSCH LFA 457 system, under constant Ar flow. The experimental uncertainties in both these measurements were 5–10%. Heat capacity measurements were performed using a NETZSCH DSC 404C system. Separate pieces of the specimens were also used for room-temperature Hall measurements and air stability tests. Air stability tests indicated that the specimens began to oxidize and degrade at 673 K, similar to that of previously reported skutterudites [13,14].

CCDC contains the supplementary crystallographic data for this paper, with deposition numbers 1562579 and 1562580 for $Yb_{0.13}In_{0.02}Co_3FeSb_{12}$ and $Yb_{0.39}In_{0.02}Co_{2.4}Fe_{1.6}Sb_{12}$, respectively. These data can be obtained free of charge via http://www.ccdc.cam.ac.uk/conts/retrieving.html (or from the CCDC, 12 Union Road, Cambridge CB2 1EZ, UK; Fax: +44 1223 336033; E-mail: deposit@ccdc.cam.ac.uk).

3. Results

3.1. Structural Characterization

Rietveld refinement profiles from the powder XRD data are shown in Figure 1, displaying calculated and observed data and the difference between them. Table 1 indicates the refinement results. The crystal structures were refined with space group $Im\overline{3}$ (#204), and the initial atomic positions were based on data from previously reported Yb filled skutterudites [11,12]. The filler Yb atoms at the $2a$ site occupy less than the nominal composition, due in part to a trace amount of Yb_2O_3 (observed in profiles at 29.7°) and to steric effects [11,12,15,16]. The Co-to-Fe ratios are extremely close to the nominal compositions. The lattice parameters are 9.0661 Å and 9.0877 Å for $Yb_{0.13}In_{0.02}Co_3FeSb_{12}$ and $Yb_{0.39}In_{0.02}Co_{2.4}Fe_{1.6}Sb_{12}$, respectively, with an increase in lattice parameter with Yb and In filling fractions and Co-to-Fe ratio, in agreement with previously reported data. Furthermore, the Yb filling fraction increases from 33% to 49% with increased Fe substitution, similarly reported for Gd and (Ba, Yb) filled skutterudites [10–17]. Elemental mapping from EDS data shows an even dispersion of elements for both specimens, indicating good homogeneity of the specimens, and corroborate our refinement results.

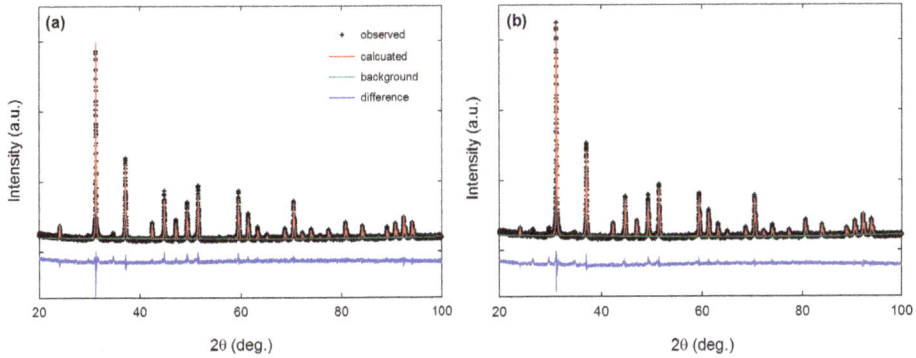

Figure 1. Powder X-ray diffraction (XRD) data for (a) $Yb_{0.13}In_{0.02}Co_3FeSb_{12}$ and (b) $Yb_{0.39}In_{0.02}Co_{2.4}Fe_{1.6}Sb_{12}$, including profile fit, profile difference, and profile residuals from Rietveld refinement.

Table 1. Rietveld refinement results for $Yb_{0.13}In_{0.02}Co_3FeSb_{12}$ and $Yb_{0.39}In_{0.02}Co_{2.4}Fe_{1.6}Sb_{12}$.

Nominal Composition	$Yb_{0.4}In_{0.02}Co_3FeSb_{12}$	$Yb_{0.8}In_{0.02}Co_{2.5}Fe_{1.5}Sb_{12}$
Composition	$Yb_{0.13}In_{0.02}Co_3FeSb_{12}$	$Yb_{0.39}In_{0.02}Co_{2.4}Fe_{1.6}Sb_{12}$
Space Group (Z)	$Im\bar{3}$ (#204), 8	
a (Å)	9.0660(6)	9.0872(8)
V (Å^3)	745.1(7)	750.4(2)
Radiation	Graphite Monochromated CuK_α (1.54056 A)	
$D_{calc.}$ (g/cm^3)	6.43	7.17
2θ range (deg.)	20–100	2–100
Step Width (deg.)	0.005	0.005
Reduced χ^2	2.40	2.84
wR_p, R_p	0.0739, 0.0581	0.0779, 0.0610
U_{iso} (Å^2) for Yb	0.0095(0)	0.0125(6)
U_{iso} (Å^2) for In	0.0090(3)	0.0101(7)
U_{iso} (Å^2) for Co/Fe	0.0070(1)	0.0039(4)
U_{iso} (Å^2) for Sb	0.0037(4)	0.0041(2)
y (Sb)	0.8436(7)	0.8423(6)
z (Sb)	0.6654(4)	0.6649(9)

Atomic Positions: Yb/In, 2a (0, 0, 0); Co/Fe, 8c (1/4, 1/4, 1/4); Sb, 24g (0, y, z).

3.2. Transport Properties

Figure 2a,b show temperature-dependent (300–800 K) S and ρ data, respectively. The $Yb_{0.13}In_{0.02}Co_3FeSb_{12}$ specimen exhibits a metallic-like temperature dependence with ρ increasing with temperature, although these values saturate to 1.6 mΩ cm^{-1} at 650 K. The ρ values for $Yb_{0.39}In_{0.02}Co_{2.4}Fe_{1.6}Sb_{12}$ do not exhibit as strong a temperature dependence in the measured temperature range. The S values for both specimens increase with increasing temperature and peak at 700 K, with values of 140 μV/K and 160 μV/K for $Yb_{0.13}In_{0.02}Co_3FeSb_{12}$ and $Yb_{0.39}In_{0.02}Co_{2.4}Fe_{1.6}Sb_{12}$, respectively. Both specimens have positive S values, indicating that holes are the majority carriers, in agreement with room-temperature Hall measurements that provide carrier concentrations (p) of 2.6×10^{20} cm^{-3} and 4×10^{20} cm^{-3} for Yb$_{0.13}$In$_{0.02}$Co$_3$FeSb$_{12}$ and Yb$_{0.39}$In$_{0.02}$Co$_{2.4}$Fe$_{1.6}$Sb$_{12}$, respectively.

In the single parabolic band model, S and p are given by [18]

$$S = \pm \frac{k_B}{e} \left(\frac{(2+r)F_{1+r}(\eta)}{(1+r)F_r(\eta)} - \eta \right) \tag{1}$$

and

$$p = \frac{4\pi(2m_e k_B T)^{3/2}}{h^3} \left(\frac{m^*}{m_e} \right)^{3/2} F_{1/2}(\eta) \tag{2}$$

where the plus and minus signs in Equation (1) are for holes (+) and electrons (−), η is the reduced Fermi energy (=$E_F/k_B T$, where E_F is the Fermi energy, k_B is the Boltzmann constant, and T is absolute temperature), F_r is the Fermi integral of order r, and r is the exponent of the energy dependence of the electron mean free path. $r = 0$ for scattering from acoustic phonons (lattice vibrations) and $r = 2$ for ionized impurity scattering. In our estimate for effective mass, m^*, the intermediate value of $r = 1$ is used. Using our room-temperature S and p values, we estimate m^* to be $0.7m_e$ for Yb$_{0.13}$In$_{0.02}$Co$_3$FeSb$_{12}$ and $1.4m_e$ for Yb$_{0.39}$In$_{0.02}$Co$_{2.4}$Fe$_{1.6}$Sb$_{12}$. These values are much smaller than that for Yb$_x$Fe$_{3.5}$Ni$_{0.5}$Sb$_{12}$ compositions, but are similar to Yb$_{0.5}$Fe$_{1.5}$Co$_{2.5}$Sb$_{12}$ and Ca$_{0.17}$Ce$_{0.05}$Fe$_{1.47}$Co$_{2.53}$Sb$_{12}$ which have a comparable Co-to-Fe content [19–21].

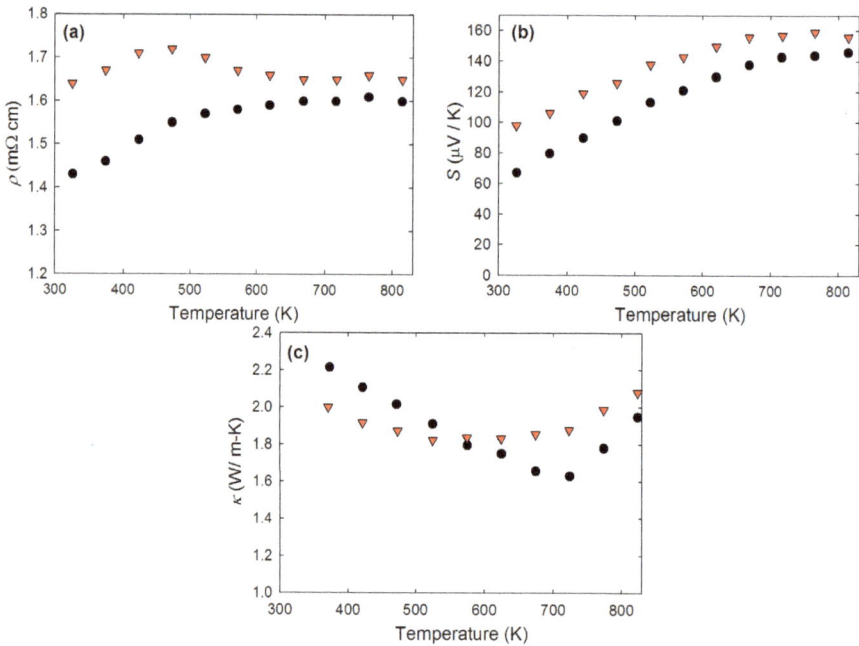

Figure 2. Temperature-dependent (**a**) ρ, (**b**) S, and (**c**) κ for Yb$_{0.13}$In$_{0.02}$Co$_3$FeSb$_{12}$ (circle) and Yb$_{0.39}$In$_{0.02}$Co$_{2.4}$Fe$_{1.6}$Sb$_{12}$ (triangle).

Figure 2c shows κ data calculated from thermal diffusivity and heat capacity measurements using the equation $\kappa = D \cdot d \cdot C_p$, where D is measured density, d is measured thermal diffusivity, and C_p is measured heat capacity. Figure 3 shows κ_L as calculated using the Wiedmann–Franz relation, where $\kappa_E = L_0 T/\rho$ (L_0 being the Lorenz number taken to be 2.45×10^{-8} V^2 K^{-2}). These κ values are smaller compared to those of previously reported (Yb, In) and (Ba, In) double-filled n-type skutterudites, as well as that of (Ce, Nd) double-filled p-type skutterudites [10,22,23]. An increase in κ and κ_L is observed above 700 K, which can be attributed to bipolar diffusion. The contribution of bipolar diffusion, κ_B, is given by $\kappa_B = \sigma_e\sigma_h(S_h-S_e)^2T/\sigma_e + \sigma_h$, where σ_e is the electron conduction, σ_h is the hole conduction, S_h is the hole Seebeck coefficient, and S_e is the electron Seebeck coefficient [1]. An estimation of κ_B can be made from the high-temperature data using $\kappa_L = 3.5(k_B/h)^3 (MV^{1/3}\theta_D{}^3/\gamma^2T)$, where h is Planck's constant, M is the average mass per atom, V is the average atomic volume, θ_D is the Debye temperature, and γ is the Grüneisen parameter. It is clear that Umklapp scattering dominates κ_L above θ_D [1]. Therefore, the inset to Figure 3 illustrates the procedure of using a fit $\kappa_L \sim T^{-1}$ to estimate κ_B for the $Yb_{0.13}In_{0.02}Co_3FeSb_{12}$ specimen, resulting in proportions of 53%, 33%, and 14% for κ_E, κ_L, and κ_B, respectively. The estimations for the $Yb_{0.39}In_{0.02}Co_{2.4}Fe_{1.6}Sb_{12}$ specimen were done with the same method, and gave values of 48%, 44%, and 8% for κ_E, κ_L, and κ_B, respectively. These κ_B values are higher than that of single-filled Yb compositions, possibly due to the additional low-lying donor states with In filling [3,10]. An increase in Fe content resulted in a near 50% reduction in κ_B [11].

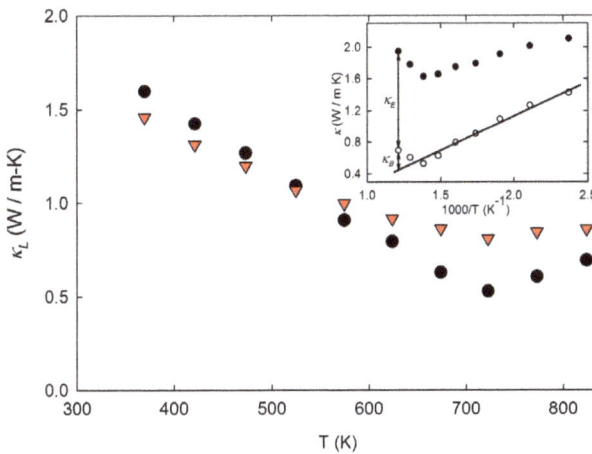

Figure 3. Temperature-dependent κ_L for $Yb_{0.13}In_{0.02}Co_3FeSb_{12}$ (circle) and $Yb_{0.39}In_{0.02}Co_{2.4}Fe_{1.6}Sb_{12}$ (triangle). The inset illustrates the method used to estimate κ_B for both specimens, with $Yb_{0.13}In_{0.02}Co_3FeSb_{12}$ shown here, where the solid line is the T^{-1} dependence between 400 K and 700 K.

Figure 4 shows the ZT values for both specimens. These values have been calculated from the measured data, and both specimens show increasing ZT with increasing temperature, with a maximum value of 0.6 at 760 K for the $Yb_{0.39}In_{0.02}Co_{2.4}Fe_{1.6}Sb_{12}$ specimen. This maximum ZT value is lower than that of the n-type (Yb, In) double-filled skutterudite but greater than (Ce, Yb) double-filled p-type skutterudites with comparable filling fraction and Co-to-Fe ratio [10,21]. However, other reported p-type (Ce, Yb) double-filled skutterudites with greater filling fraction and Fe substitution exhibit a larger ZT (=0.87) [24].

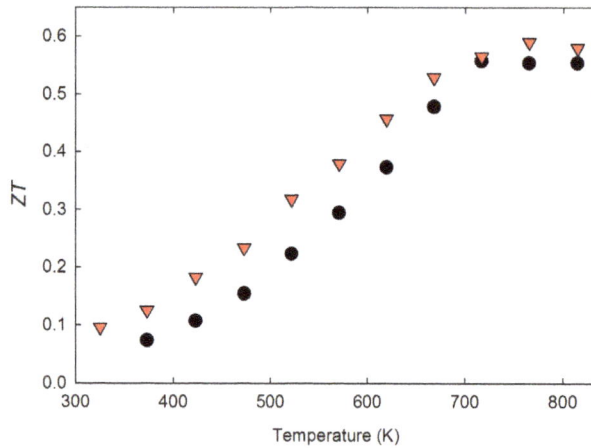

Figure 4. Temperature-dependent ZT for $Yb_{0.13}In_{0.02}Co_3FeSb_{12}$ (circle) and $Yb_{0.39}In_{0.02}Co_{2.4}Fe_{1.6}Sb_{12}$ (triangle).

4. Conclusions

The structural and high-temperature transport properties of p-type (Yb, In) double-filled skutterudites were investigated. We observed an increase in both ρ and S for $Yb_{0.39}In_{0.02}Co_{2.4}Fe_{1.6}Sb_{12}$, whereas κ was less for the specimen with lower Yb content. Above 700 K, both specimens exhibited a fairly large κ_B contribution that significantly increased κ above this temperature, although κ_B decreased with increasing Fe content. Both specimens exhibited the largest ZT values at 750 K. The performance of these p-type skutterudites may be enhanced by a further increase in overall filling fractions of (Yb, In), corresponding to an increase in Fe content.

Acknowledgments: This work was supported by the II-VI Foundation Block-Gift Program. The authors thank Jeff Sharp of Marlow Industries for air stability testing.

Author Contributions: George S. Nolas conceived and designed the experiments; Dean Hobbis and Kaya Wei performed synthesis, densification and preparation of specimens; Yamei Liu and Terry M. Tritt performed high temperature transport measurements on specimens; Dean Hobbis, Kaya Wei and George S. Nolas analyzed the data; Dean Hobbis wrote the manuscript. All authors contributed to the experiment, the analysis of the data, and edition of the manuscript.

Conflicts of Interest: The authors declare no conflict of interest.

References

1. Nolas, G.S.; Sharp, J.W.; Goldsmid, H.J. *Thermoelectrics: Basic Principles and New Material Developments*; Springer: Berlin, Germany, 2001.
2. Stabler, F.R. Commercialization of Thermoelectric Technology. *Mater. Res. Symp. Proc.* **2006**, *886*, 13–21. [CrossRef]
3. Uher, C. Skutterudites: Prospective novel thermoelectrics. In *Semiconductors and Semimetals*; Tritt, T.M., Ed.; Academic Press: San Diego, CA, USA, 2001; Volume 69, pp. 139–253; ISSN 978-0-12-752178-7.
4. Dahal, T.; Kim, H.S.; Gahlawat, S.; Dahal, K.; Jie, Q.; Liu, W.; Lan, Y.; White, K.; Ren, Z. Transport and mechanical properties of the double-filled p-type skutterudites $La_{0.68}Ce_{0.22}Fe_{4-x}Co_xSb_{12}$. *Acta Mater.* **2016**, *117*, 13–22. [CrossRef]
5. Salvador, J.R.; Yang, J.; Shi, X.; Wang, H.; Wereszczak, A.A.; Kong, H.; Uher, C. Transport and mechanical properties of Yb-filled skutterudites. *Philos. Mag.* **2009**, *89*, 1517–1534. [CrossRef]

6. Zhang, L.; Rogl, G.; Grytsiv, A.; Puchegger, S.; Koppensteiner, J.; Spieckermann, F.; Kabelka, H.; Reinecker, M.; Rogl, P.; Schranz, W.; et al. Mechanical properties of filled antimonide skutterudites. *Mater. Sci. Eng. B* **2010**, *170*, 26–31. [CrossRef]

7. Nolas, G.S.; Fowler, G. Partial filling of skutterudites: Optimization for thermoelectric applications. *J. Mater. Res.* **2005**, *20*, 3234–3237. [CrossRef]

8. Nolas, G.S.; Morelli, D.T.; Tritt, T.M. Skutterudites: A phonon-glass-electron-crystal approach to advanced thermoelectric energy conversion applications. *Annu. Rev. Mater. Res. Bull.* **1999**, *29*, 199–205. [CrossRef]

9. Nolas, G.S.; Kaeser, M.; Littleton, R.T.; Tritt, T.M. High figure of merit in partially filled ytterbium skutterudite materials. *Appl. Phys. Lett.* **2000**, *77*, 1855. [CrossRef]

10. Peng, J.Y.; Alboni, P.N.; He, J.; Zhang, B.; Su, Z.; Holgate, T.; Gothard, N.; Tritt, T.M. Thermoelectric properties of (In,Yb) double-filled $CoSb_3$ skutterudite. *J. Appl. Phys.* **2008**, *104*, 053710. [CrossRef]

11. Dong, Y.; Puneet, P.; Tritt, T.M.; Nolas, G.S. Crystal structure and high temperature transport properties of Yb-filled p-type skutterudites $Yb_xCo_{2.5}Fe_{1.5}Sb_{12}$. *J. Solid State Chem.* **2014**, *209*, 1–5. [CrossRef]

12. Dong, Y.; Puneet, P.; Tritt, T.M.; Nolas, G.S. High temperature thermoelectric properties of p-type skutterudites $Yb_xCo_3FeSb_{12}$. *Phys. Status Solidi RRL* **2013**, *7*, 418–420. [CrossRef]

13. Zhao, D.; Tian, C.; Tang, S.; Liu, Y.; Chen, L.D. High temperature oxidation behavior of cobalt triantimonide thermoelectric material. *J. Alloys Compd.* **2010**, *504*, 552–558. [CrossRef]

14. Shin, D.K.; Kim, I.H.; Park, K.H.; Lee, S.; Seo, W.S. Thermal Stability of $La_{0.9}Fe_3CoSb_{12}$. *J. Electron. Mater.* **2015**, *44*, 1858–1863. [CrossRef]

15. Dong, Y.; Puneet, P.; Tritt, T.M.; Martin, J.; Nolas, G.S. High temperature thermoelectric properties of p-type skutterudites $Ba_xYb_yCo_{4-z}Fe_zSb_{12}$. *J. Appl. Phys.* **2012**, *112*, 083718. [CrossRef]

16. Liu, R.; Chen, X.; Qiu, P.; Liu, J.; Yang, J.; Huang, X.; Chen, L. Low thermal conductivity and enhanced thermoelectric performance of Gd-filled skutterudites. *J. Appl. Phys.* **2011**, *109*, 023719. [CrossRef]

17. Zhou, C.; Morelli, D.; Zhou, X.; Wang, G.; Uher, C. Thermoelectric properties of P-type Yb-filled skutterudite YbxFeyCo4-ySb12. *Intermetallics* **2011**, *19*, 1390–1393. [CrossRef]

18. Slack, G.A.; Hussain, M.A. The maximum possible conversion efficiency of silicon-germanium thermoelectric generators. *J. Appl. Phys.* **1991**, *70*, 2694. [CrossRef]

19. Cho, J.Y.; Ye, Z.; Tessema, M.M.; Waldo, R.A.; Salvador, J.R.; Yang, J.; Cai, W.; Wang, H. Thermoelectric properties of p-type skutterudites YbxFe3.5Ni0.5Sb12 (0.8 < x < 1). *Acta Mater.* **2012**, *60*, 2104–2110.

20. Tang, X.; Li, H.; Zhang, Q.; Niino, M.; Goto, T. Synthesis and thermoelectric properties of double-atom-filled skutterudite compounds $Ca_mCe_nFe_xCo_{4-x}Sb_{12}$. *J. Appl. Phys.* **2006**, *100*, 123702. [CrossRef]

21. Yang, K.; Cheng, H.; Hng, H.H.; Ma, J.; Mi, J.L.; Zhao, X.B.; Zhu, T.J.; Zhang, Y.B. Synthesis and thermoelectric properties of double-filled skutterudites $Ce_yYb_{0.5-y}Fe_{1.5}Co_{2.5}Sb_{12}$. *J. Alloys Compd.* **2009**, *467*, 528–532. [CrossRef]

22. Zhao, W.; Wei, P.; Zhang, Q.; Dong, C.; Liu, L.; Tang, X. Enhanced Thermoelectric Performance in Barium and Indium Double-Filled Skutterudite Bulk Materials via Orbital Hybridization Induced by Indium Filler. *J. Amer. Chem. Soc.* **2009**, *131*, 3713–3720. [CrossRef] [PubMed]

23. Jie, Q.; Wang, H.; Liu, W.; Wang, H.; Chen, G.; Ren, Z. Fast Phase formation of double-filled p-type skutterudites by ball-milling and hot-pressing. *Phys. Chem. Chem. Phys.* **2013**, *15*, 6809–6816. [CrossRef] [PubMed]

24. Joo, G.S.; Shin, D.K.; Kim, I.H. Synthesis and Thermoelectric Properties of p-Type Double-Filled Ce1-zYbzFe4-xCoxSb12 Skutterudites. *J. Electron. Mater.* **2016**, *45*, 1251–1256. [CrossRef]

© 2017 by the authors. Licensee MDPI, Basel, Switzerland. This article is an open access article distributed under the terms and conditions of the Creative Commons Attribution (CC BY) license (http://creativecommons.org/licenses/by/4.0/).

crystals

MDPI

Article

Microstructure Analysis and Thermoelectric Properties of Melt-Spun Bi-Sb-Te Compounds

Weon Ho Shin [1], Jeong Seop Yoon [1], Mahn Jeong [1], Jae Min Song [1], Seyun Kim [2],
Jong Wook Roh [2], Soonil Lee [1], Won Seon Seo [1], Sung Wng Kim [3,*] and Kyu Hyoung Lee [4,*]

[1] Energy Materials Center, Energy & Environment Division, Korea Institute of Ceramic
     Engineering & Technology, Jinju 52851, Korea; whshin@kicet.re.kr (W.H.S.); yujsyoon@kicet.re.kr (J.S.Y.);
     sensjm@kicet.re.kr (M.J.); love8767@kicet.re.kr (J.M.S.); silee@kicet.re.kr (S.L.); wsseo@kicet.re.kr (W.S.S.)
[2] Materials R & D Center, Samsung Advanced Institute of Technology, Samsung Electronics, Suwon 16419,
     Korea; seyuni.kim@samsung.com (S.K.); jw.roh@samsung.com (J.W.R.)
[3] Department of Energy Science, Sungkyunkwan University, Suwon 16419, Korea
[4] Department of Nano Applied Engineering, Kangwon National University, Chuncheon 24341, Korea
*   Correspondence: kimsungwng@skku.edu (S.W.K.); khlee2014@kangwon.ac.kr (K.H.L.);
    Tel.: +82-31-299-6274 (S.W.K.); +82-33-250-6261 (K.H.L.)

Academic Editor: George S. Nolas
Received: 25 May 2017; Accepted: 19 June 2017; Published: 20 June 2017

Abstract: In order to realize high-performance thermoelectric materials, a way to obtain small grain size is necessary for intensification of the phonon scattering. Here, we use a melt-spinning-spark plasma sintering process for making p-type $Bi_{0.36}Sb_{1.64}Te_3$ thermoelectric materials and evaluate the relation between the process conditions and thermoelectric performance. We vary the Cu wheel rotation speed from 1000 rpm (~13 ms^{-1}) to 4000 rpm (~52 ms^{-1}) during the melt spinning process to change the cooling rate, allowing us to control the characteristic size of nanostructure in melt-spun $Bi_{0.36}Sb_{1.64}Te_3$ ribbons. The higher wheel rotation speed decreases the size of nanostructure, but the grain sizes of sintered pellets are inversely proportional to the nanostructure size after the same sintering condition. As a result, the ZT values of the bulks fabricated from 1000–3000 rpm melt-spun ribbons are comparable each other, while the ZT value of the bulk from the 4000 rpm melt-spun ribbons is rather lower due to reduction of grain boundary phonon scattering. In this work, we can conclude that the smaller nanostructure in the melt spinning process does not always guarantee high-performance thermoelectric bulks, and an adequate following sintering process must be included.

Keywords: thermoelectric; phonon scattering; melt spinning; $Bi_{0.36}Sb_{1.64}Te_3$; nanostructure

1. Introduction

The increasing world-wide demands on new technology for CO_2 reduction and global warming have induced the development of highly-efficient energy harvesting technology that reuses exhausted energy and shifts from fossil fuels to renewable energy and replaces it with new energy sources. Currently, more than 60% of primary energy used in industry or in combustion engines is lost as waste heat. Thermoelectricity is considered to be one of the encouraging energy harvesting technologies in the view of changing waste heat into electricity in the semiconductor materials [1,2]. The efficiency of thermoelectric (TE) devices is highly related with the performance of TE materials, which is determined by a dimensionless figure of merit, ZT, calculated by $ZT = \sigma S^2 T / \kappa$, where σ, S, T, and κ are the electrical conductivity, Seebeck coefficient, absolute temperature, and thermal conductivity, respectively [3,4]. High σ, high S, and low κ are essential to make high-performance TE materials; however, these transport parameters are correlated with each other in terms of carrier concentration, practically limiting to the ZT value of ~1.

Bi_2Te_3-based alloys are considered to be the best TE materials for near room temperature [5,6], and are the only commercialized materials applied for TE cooling and low-temperature power generation applications; however, they are still limited in special fields due to a low TE performance [7,8]. In order to design high TE materials, two different approaches are proposed: (1) enhancing power factor (PF = σS^2) and (2) reducing κ. The κ is divided by two terms of electronic contribution (κ_{ele}) and lattice contribution ($\kappa_{lat} = \kappa - \kappa_{ele}$). The κ_{ele} is in proportion to σ according to the Wiedemann–Franz law $\kappa_{ele} = L\sigma T$, where L is the Lorenz number [9]. In this context, many researchers are focusing on the κ_{lat} reduction which is considered as an independent variable to enhance TE performance [1,8,10,11]. Recent studies have shown a significant reduction of κ_{lat} with integrated defects, resulting in high ZT values for various TE materials [8,12–14]. However, these technologies are still far from commercialization in terms of mass production. The continuous attempts to reduce grain size of the TE materials have been tried via high-energy ball milling [15,16], spark erosion [17], melt spinning (MS) [8,18–20], and bottom-up chemical synthesis processes [21–23].

According to several reports [8,18–20], the MS process is proven to be effective for the realization of nano-scale sized Bi_2Te_3-based TE materials. However, it is necessary to investigate the process variables, leading to different microstructures and transport properties which are critical factors for TE performance. Adjusting the wheel rotation speed during the MS process can induce the variation of thickness and microstructure of melt-spun ribbons, which control the TE performance of their bulks with the same composition. In the present work, we have investigated the change of microstructure and TE properties of polycrystalline bulks of $Bi_{0.36}Sb_{1.64}Te_3$ fabricated by combined technique of MS and spark plasma sintering (SPS) in an effort to optimize the process parameters for nanograin structured TE materials.

2. Experimental Details

High purity elemental Bi (99.999%, 5N Plus), Sb (99.999%, 5N Plus), and Te (99.999%, 5N Plus) granules as starting materials were weighed according to the formula of $Bi_{0.36}Sb_{1.64}Te_3$. Excess 1 wt % Te was added due to Te evaporation. The raw materials were loaded into a vacuum-sealed quartz ampule and then melted at 1373 K for 4 h. The obtained ingots were pulverized, and the powders were compacted by a hydraulic press. The compactions were put into a graphite nozzle with 0.4 mm diameter. The MS process was used for the fabrication of nanostructured ribbons of $Bi_{0.36}Sb_{1.64}Te_3$. The Cu wheel (diameter ~250 mm) rotation speeds were varied as 1000 rpm (~13.1 ms^{-1}), 2000 rpm (~26.2 ms^{-1}), 3000 rpm (~39.3 ms^{-1}), and 4000 rpm (~52.4 ms^{-1}). Thin ribbons were produced by MS process, and the ribbons show amorphous structure on the contact surface and crystalline nanostructure on the free surface. The melt-spun ribbons were pulverized into powders and sintered using SPS at 753 K for 3 min under 60 MPa. The dimension of SPSed pellets is in the diameter of 12.5 mm and the height of 10 mm in this work.

X-ray diffraction (XRD, New D8 Advance, Bruker, Cu Kα) for SPSed pellets was performed on parallel and perpendicular planes of SPS pressing direction. The microstructure was investigated using scanning electron microscopy (SEM) (JSM-7600F, JEOL, Peabody, MA, USA). The temperature dependences of the σ and S were measured from room temperature to 473 K by a four-point probe method using the ZEM-3 apparatus (ULVAC-RIKO). Carrier concentrations and mobilities were obtained from Hall effect measurement system (HT-Hall, ResiTest 8300, Toyo Corporation, Toyo, Japan). The κ values were measured by laser-flash analysis (LFA) using TA Netzsch LFA 457. All measured TE transport data were acquired at the same dimension and configuration, and were obtained within the experimental error of σ (~4%), S (~4%), and κ (~5%). Thus, we assume total uncertainty of ZT as ~12%.

3. Results and Discussion

3.1. Microstructure Analysis

After the MS process, we can obtain thin and short (~2 mm in width and ~10 mm in length) ribbon-shaped materials with an amorphous structure on the contact surface and nano-scale structure on the free surface [20]. The Cu wheel rotation speed during the MS process would be a major factor determining the cooling rate. Figure 1 shows the SEM images of the cross-section of melt-spun $Bi_{0.36}Sb_{1.56}Te_3$ ribbons with varying wheel rotation speeds of 1000 rpm, 2000 rpm, and 4000 rpm. As shown in Figure 1, increasing wheel rotation speed from 1000 rpm to 4000 rpm results in a decrease of the ribbon thickness from 17.8 μm to 3.38 μm. The thickness is inversely proportional with the wheel rotation speed, which is in good agreement with the previous reports [24,25], and the melt-spun ribbon is composed of non-crystalline contact surface and crystalline free surface [20]. The nanostructure of the free surface also changes with the wheel rotation speed, as shown in Figure 2. The thickness of rod-shaped nanograin decreased with increasing wheel rotation speed: 503 nm for 1000 rpm, 451 nm for 2000 rpm, and 372 nm for 4000 rpm, respectively, averaged by more than 20 batches of melt-spun ribbons. It is also noteworthy that the pore sizes among nanostructures were smaller as the wheel rotation speed increased, caused by higher growth rate. We assume that this difference in the characteristic sizes of nanostructures in melt-spun ribbons could give a significant difference in the TE performance of SPSed pellets.

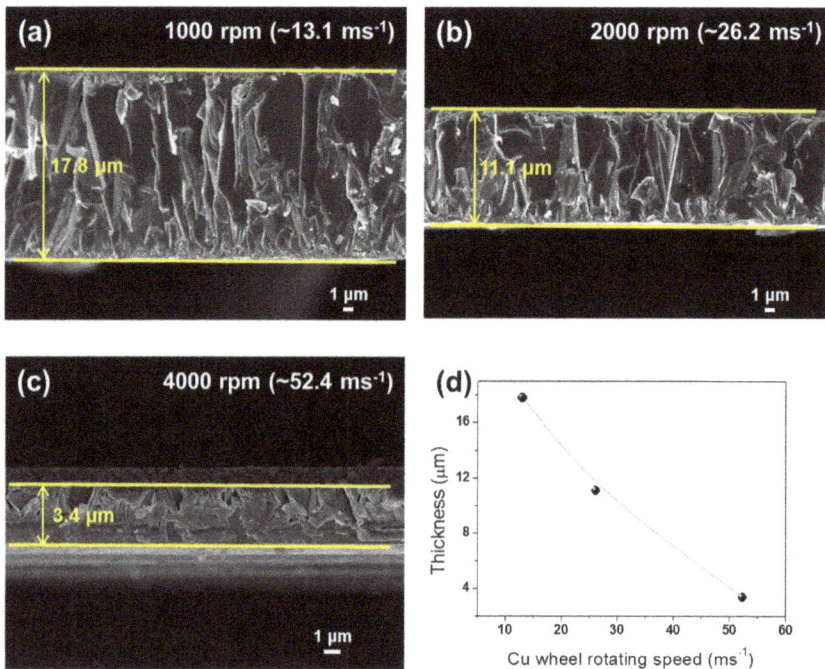

Figure 1. The cross-section SEM images of the melt-spun ribbons with varying Cu wheel rotation speed during melt spinning (MS) process. (**a**) 1000 rpm; (**b**) 2000 rpm; (**c**) 4000 rpm and (**d**) Plot of thickness with Cu wheel rotation speed (ms^{-1}).

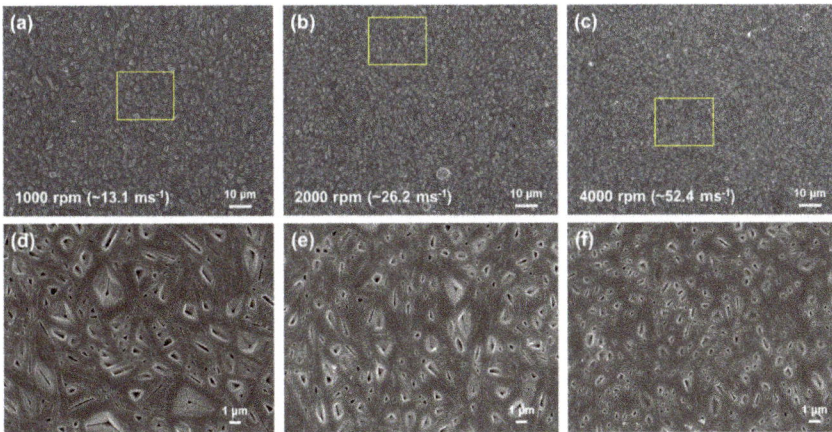

Figure 2. The SEM images of the free surfaces of ribbons with varying wheel rotation speed during the MS process: (**a,d**) 1000 rpm; (**b,e**) 2000 rpm; (**c,f**) 4000 rpm.

The melt-spun ribbons were ground into powders by using a mortar and sintered by using SPS technique. The relative densities of all the samples were >98%. Figure 3 shows the X-ray diffraction patterns of the SPSed pellets of $Bi_{0.36}Sb_{1.64}Te_3$ fabricated from melt-spun ribbons at 1000 (BST1000), 2000 (BST2000), and 4000 (BST4000) rpm wheel rotation speed. All samples show the fundamental diffraction peaks from Sb_2Te_3 with rhombohedral structure (JCPDS # 65-3678, R-3m space group). When the wheel speed was more than 2000 rpm, we can see an additional peak at $2\theta = 27.6°$ corresponding to Te secondary phase (JCPDS # 36-1452). Meanwhile, the higher cooling rate led to the production of a larger amount of Te secondary phase, which can be easily found in Bi-Te-based TE materials [15]. There is no large difference of peak shift or peak intensity, which means that the structure of $Bi_{0.36}Sb_{1.64}Te_3$ is unchangeable with varying wheel rotation speed from 1000 rpm to 4000 rpm during the MS process. The fractured surfaces of the SPSed bulks are shown in Figure 4 to investigate the microstructure evolution during the SPS process. Interestingly, the grain size increased with increase of wheel rotation speed in the range of 1000 rpm to 4000 rpm. We assumed that the higher MS wheel rotation speed generates a smaller nanostructure of melt-spun ribbon, and thus the smaller nanograin structure can be maintained after the SPS process. However, our experimental results do not correspond with this assumption. As shown in Figure 4, the grain sizes of the SPSed pellets of BST1000, BST2000, and BST4000 are 17.9 μm, 19.4 μm, and 20.5 μm, respectively. It is noted that the grain sizes of SPSed pellets are much larger than the characteristic sizes of melt-spun ribbons. This phenomenon can be explained by the grain growth kinetics of nanoparticles. The smaller nanoparticles have lower melting point due to enhanced surface diffusion rate, leading to enlarged grain size [26]. Additionally, a larger amount of amorphous phase at higher wheel rotation speed can also be the origin of different grain size in the same sintering process [27]. This result suggests that a precisely controlled ultra-fast sintering process is required to obtain nanograin structured bulks from melt-spun ribbons.

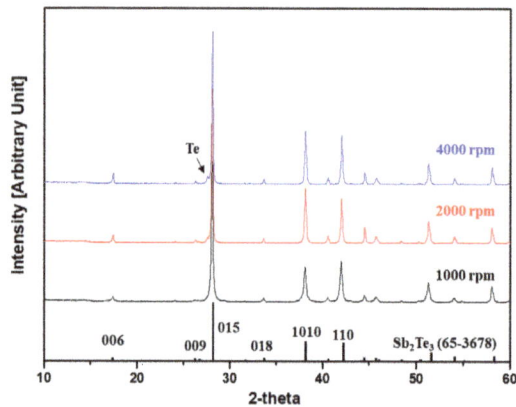

Figure 3. The XRD patterns of polycrystalline bulks of $Bi_{0.36}Sb_{1.64}Te_3$ fabricated from melt-spun ribbons at 1000, 2000, and 4000 rpm wheel rotation speed during the MS process. The bottom is the typical peak of Sb_2Te_3 (JCPDS # 65-3678).

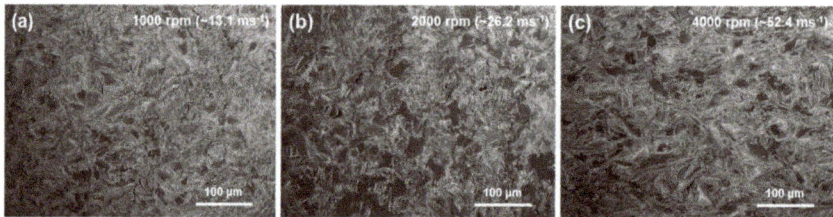

Figure 4. SEM images of fractured surfaces of the bulks of $Bi_{0.36}Sb_{1.64}Te_3$ fabricated from melt-spun ribbons at (**a**) 1000 (BST1000), (**b**) 2000 (BST2000), and (**c**) 4000 (BST4000) rpm wheel rotation speed during the MS process.

3.2. Thermoelectric Properties

Figure 5 shows temperature-dependent σ values for BST1000–BST4000 samples. All samples showed a decrease of σ with increasing temperature in the measured temperature range, indicating a metallic behavior. For BST1000, the room temperature σ value was 955 Scm^{-1}, while BST2000–BST4000 samples had similar room-temperature σ values of 1192 Scm^{-1}, 1201 Scm^{-1}, and 1163 Scm^{-1}, respectively. To clarify the electronic transport mechanism, the carrier concentration values of the samples were calculated by measuring Hall coefficient (R_H), as the following equation:

$$R_H = 1/pe \tag{1}$$

where p and e correspond to hole carrier concentration and electron charge, respectively. The carrier mobility (μ) is calculated by the following equation:

$$\sigma = pe\mu \tag{2}$$

Figure 5b depicts the room-temperature carrier concentration values for BST1000–BST4000 samples, which ranged from 2.87×10^{19} cm^{-3} to 3.86×10^{19} cm^{-3}. Interesting behavior was observed in the carrier mobility as shown in Figure 5c, in which the mobility increased from 191 $cm^2 \cdot V^{-1} \cdot s^{-1}$ to

254 cm^2·V^{-1}·s^{-1} with increase in the wheel rotation speed. This result is due to the reduced carrier scattering originating from the larger grain size at higher wheel rotation speed, as shown in Figure 4.

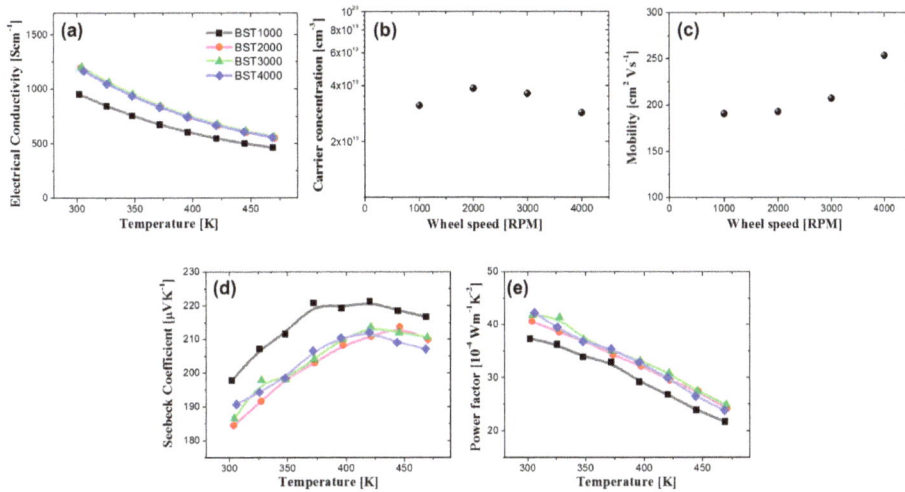

Figure 5. (**a**) Temperature dependence of electrical conductivity (σ); (**b**) carrier concentration and (**c**) mobility at room temperature; (**d**) temperature dependence of Seebeck coefficient (*S*); (**e**) temperature dependence of power factor (PF) for BST1000–BST4000 samples.

Figure 5d shows the temperature dependence of *S* for BST1000–BST4000 samples. The *S* values for all samples are positive, indicating that the major charge carrier is hole. The *S* increased with increasing temperature, reached a maximum at ~400 K, and decreased at high temperature, which is typical behavior presented in Bi$_2$Te$_3$-based TE materials mainly originating from the thermal excitation of minority carrier. The room temperature *S* values are 198 µV·K^{-1} for BST1000, 184 µV·K^{-1} for BST2000, 187 µV·K^{-1} for BST3000, and 191 µV·K^{-1} for BST4000, respectively. The *S* values of BST1000 showed the highest value within the measured temperature range due to the lower carrier concentration. Figure 5e shows the temperature dependences of PF values for BST1000–BST4000 samples. The PF value for BST1000 shows the lowest value of 37 × 10^{-4} Wm^{-1}·K^{-2} at room temperature due to the lowest σ, while the PF values of BST2000–BST4000 samples were almost the same over the whole measure temperature range (PF = 41–42 × 10^{-4} Wm^{-1}·K^{-2} at room temperature).

The temperature dependence of κ is displayed in Figure 6a. All samples showed similar temperature dependency of κ, where it decreased at low temperature and increased at high temperature, suggesting the effect of bipolar thermal conduction, while κ value increased with wheel rotation speed. To clarify this, we calculated the κ_{lat} value by subtraction of κ_{ele} from κ, in which κ_{ele} was estimated by the Wiedemann–Franz law ($\kappa_{ele} = L\sigma T$), using $L = 2.0 \times 10^{-8}$ V^2·K^{-2} for degenerate semiconductor [28,29]. The κ_{lat} calculated from the above equations is shown in Figure 6b. The κ_{lat} of BST4000 was >10% higher than those of other samples, and κ_{lat} of BST2000 showed the lowest value within whole measured temperature range. These results fit well with the grain size variation against wheel rotation speed, which was already described by the microstructure of SPSed bulks. The largest grain size of BST4000 led to a reduction in the grain boundary phonon scattering, resulting in increased κ_{lat}.

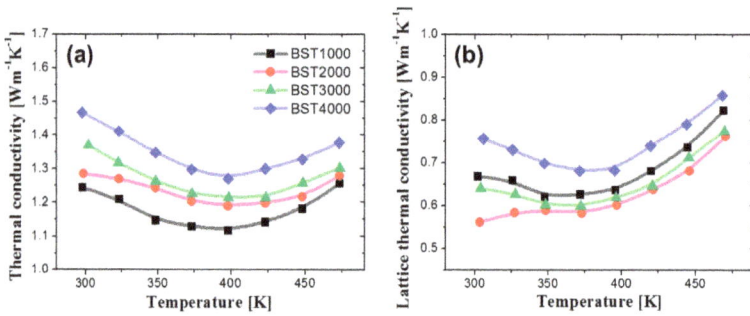

Figure 6. Temperature dependence of (**a**) total thermal conductivity (κ), and (**b**) lattice thermal conductivity (κ_{lat}) for BST1000–BST4000 samples.

The dimensionless figure of merit (ZT) values as a function of temperature for BST1000–BST4000 samples are shown in Figure 7. The ZT value increased with the temperature, and after a maximum near 400 K, it decreased at higher temperatures. The overall ZT values did not change significantly with the wheel rotation speed within the instrumental error range (~12%); the maximum ZT values were ~1.08 @ 370 K for BST1000, ~1.07 @ 400 K for BST2000, ~1.09 @ 400 K for BST3000, and ~1.02 @ 400 K for BST4000, respectively. Interestingly, BST4000 showed the lowest ZT_{max} compared to other SPSed bulk samples (BST1000–BST3000) despite the smallest characteristic size of its melt-spun ribbon (Figure 2). This is considered to be related to the microstructural evolution during the sintering process. As shown in Figure 2, the higher wheel rotation speed resulted in the smaller characteristic nanostructure size in melt-spun ribbons; however, the nano-scale grain structure could not be maintained in SPSed bulks due to the rapid grain growth during the SPS process, and grain growth was more prominent for BST4000. Therefore, some other fast sintering approaches to maintaining the nanostructure in melt-spun ribbon are highly required for boosting the grain boundary phonon scattering and ZT value.

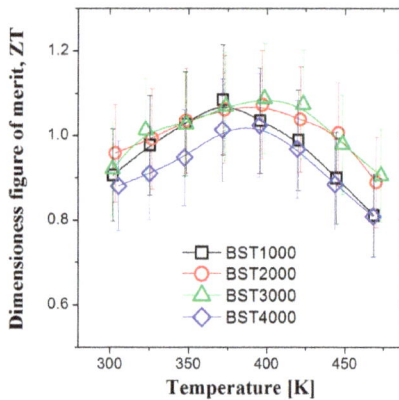

Figure 7. Temperature dependence of the dimensionless figure of merit (ZT) for BST1000–BST4000 samples.

4. Conclusions

We have synthesized polycrystalline bulks of $Bi_{0.36}Sb_{1.64}Te_3$ using melt spinning and spark plasma sintering processes. We varied the wheel rotation speed between 1000 rpm and 4000 rpm during the MS process, and the microstructures of the melt spun ribbons and sintered pellets were investigated. The thickness of the ribbon and the characteristic size of nanostructure on the free surface were reduced as the wheel rotation speed increased due to higher cooling rate, while the grain size after spark plasma sintering significantly increased in value of ~20 μm for all SPSed bulk samples due to the rapid grain growth. The *ZT* values of BST1000–BST4000 did not change significantly with Cu wheel rotation speed, suggesting that small nanostructure size in precursors is not always preferable to enhance the thermoelectric performance of Bi_2Te_3-based TE materials in melt spinning process.

Acknowledgments: This work was supported by Dual Use Technology Program (12-DU-MP-01) of Agency for Defense Development, Republic of Korea, and also supported by a National Research Foundation of Korea (NRF) grant funded by the Korean Government (MSIP) (NRF-2015R1A5A1036133).

Author Contributions: W.H.S. and K.H.L. conceived and designed the experiments; J.S.Y., J.M., and J.M.S. performed the experiments; S.K. and R.J.W. analyzed the data; S.L. and W.S.S. contributed reagents/materials/analysis tools; W.H.S., S.W.K., and K.H.L. wrote the paper. All the authors contributed to the experiments, the analysis of the data, and edition of the manuscript.

Conflicts of Interest: The authors declare no conflict of interest.

References

1. Snyder, G.J.; Toberer, E.S. Complex thermoelectric materials. *Nat. Mater.* **2008**, *7*, 105–114. [CrossRef] [PubMed]
2. Bell, L.E. Cooling, heating, generating power, and recovering waste heat with thermoelectric systems. *Science* **2008**, *321*, 1457–1461. [CrossRef] [PubMed]
3. DiSalvo, F.J. Thermoelectric cooling and power generation. *Science* **1999**, *285*, 703–706. [CrossRef] [PubMed]
4. Heremans, J.P.; Jovovic, V.; Toberer, E.S.; Saramat, A.; Kurosaki, K.; Charoenphakdee, A.; Yamanaka, S.; Snyder, G.J. Enhancement of thermoelectric efficiency in pbte by distortion of the electronic density of states. *Science* **2008**, *321*, 554–557. [CrossRef] [PubMed]
5. Goldsmid, H. Bismuth telluride and its alloys as materials for thermoelectric generation. *Materials* **2014**, *7*, 2577–2592. [CrossRef]
6. Lee, K.H.; Kim, S.W. Design and preparation of high-performance bulk thermoelectric materials with defect structures. *J. Korean Ceram. Soc.* **2017**, *54*, 75–85. [CrossRef]
7. Wood, C. Materials for thermoelectric energy conversion. *Rep. Prog. Phys.* **1988**, *51*, 459. [CrossRef]
8. Kim, S.I.; Lee, K.H.; Mun, H.A.; Kim, H.S.; Hwang, S.W.; Roh, J.W.; Yang, D.J.; Shin, W.H.; Li, X.S.; Lee, Y.H. Dense dislocation arrays embedded in grain boundaries for high-performance bulk thermoelectrics. *Science* **2015**, *348*, 109–114. [CrossRef] [PubMed]
9. Kim, H.-S.; Gibbs, Z.M.; Tang, Y.; Wang, H.; Snyder, G.J. Characterization of lorenz number with seebeck coefficient measurement. *APL Mater.* **2015**, *3*, 041506. [CrossRef]
10. Kim, W.; Zide, J.; Gossard, A.; Klenov, D.; Stemmer, S.; Shakouri, A.; Majumdar, A. Thermal conductivity reduction and thermoelectric figure of merit increase by embedding nanoparticles in crystalline semiconductors. *Phys. Rev. Lett.* **2006**, *96*, 045901. [CrossRef] [PubMed]
11. Poudel, B.; Hao, Q.; Ma, Y.; Lan, Y.; Minnich, A.; Yu, B.; Yan, X.; Wang, D.; Muto, A.; Vashaee, D. High-thermoelectric performance of nanostructured bismuth antimony telluride bulk alloys. *Science* **2008**, *320*, 634–638. [CrossRef] [PubMed]
12. Tan, G.; Shi, F.; Hao, S.; Zhao, L.-D.; Chi, H.; Zhang, X.; Uher, C.; Wolverton, C.; Dravid, V.P.; Kanatzidis, M.G. Non-equilibrium processing leads to record high thermoelectric figure of merit in PbTe-SrTe. *Nat. Commun.* **2016**, *7*, 12167. [CrossRef] [PubMed]
13. Zhao, L.-D.; Lo, S.-H.; Zhang, Y.; Sun, H.; Tan, G.; Uher, C.; Wolverton, C.; Dravid, V.P.; Kanatzidis, M.G. Ultralow thermal conductivity and high thermoelectric figure of merit in snse crystals. *Nature* **2014**, *508*, 373–377. [CrossRef] [PubMed]
14. Rhyee, J.-S.; Lee, K.H.; Lee, S.M.; Cho, E.; Kim, S.I.; Lee, E.; Kwon, Y.S.; Shim, J.H.; Kotliar, G. Peierls distortion as a route to high thermoelectric performance in $In_4Se_{3-\delta}$ crystals. *Nature* **2009**, *459*, 965–968. [CrossRef] [PubMed]
15. Lan, Y.; Poudel, B.; Ma, Y.; Wang, D.; Dresselhaus, M.S.; Chen, G.; Ren, Z. Structure study of bulk nanograined thermoelectric bismuth antimony telluride. *Nano Lett.* **2009**, *9*, 1419–1422. [CrossRef] [PubMed]

16. Ma, Y.; Hao, Q.; Poudel, B.; Lan, Y.; Yu, B.; Wang, D.; Chen, G.; Ren, Z. Enhanced thermoelectric figure-of-merit in p-type nanostructured bismuth antimony tellurium alloys made from elemental chunks. *Nano Lett.* **2008**, *8*, 2580–2584. [CrossRef] [PubMed]

17. Nguyen, P.K.; Lee, K.H.; Moon, J.; Kim, S.I.; Ahn, K.A.; Chen, L.H.; Lee, S.M.; Chen, R.K.; Jin, S.; Berkowitz, A.E. Spark erosion: A high production rate method for producing $Bi_{0.5}Sb_{1.5}Te_3$ nanoparticles with enhanced thermoelectric performance. *Nanotechnology* **2012**, *23*, 415604. [CrossRef] [PubMed]

18. Cai, X.; Fan, X.A.; Rong, Z.; Yang, F.; Gan, Z.; Li, G. Improved thermoelectric properties of $Bi_2Te_{3-x}Se_x$ alloys by melt spinning and resistance pressing sintering. *J. Phys. D Appl. Phys.* **2014**, *47*, 115101. [CrossRef]

19. Ebling, D.G.; Jacquot, A.; Jägle, M.; Böttner, H.; Kühn, U.; Kirste, L. Structure and thermoelectric properties of nanocomposite bismuth telluride prepared by melt spinning or by partially alloying with IV–VI compounds. *Phys. Status Solidi RRL* **2007**, *1*, 238–240. [CrossRef]

20. Xie, W.; He, J.; Kang, H.J.; Tang, X.; Zhu, S.; Laver, M.; Wang, S.; Copley, J.R.; Brown, C.M.; Zhang, Q. Identifying the specific nanostructures responsible for the high thermoelectric performance of $(Bi,Sb)_2Te_3$ nanocomposites. *Nano Lett.* **2010**, *10*, 3283–3289. [CrossRef] [PubMed]

21. Min, Y.; Park, G.; Kim, B.; Giri, A.; Zeng, J.; Roh, J.W.; Kim, S.I.; Lee, K.H.; Jeong, U. Synthesis of multishell nanoplates by consecutive epitaxial growth of Bi_2Se_3 and Bi_2Te_3 nanoplates and enhanced thermoelectric properties. *ACS Nano* **2015**, *9*, 6843–6853. [CrossRef] [PubMed]

22. Min, Y.; Roh, J.W.; Yang, H.; Park, M.; Kim, S.I.; Hwang, S.; Lee, S.M.; Lee, K.H.; Jeong, U. Surfactant-free scalable synthesis of Bi_2Te_3 and Bi_2Se_3 nanoflakes and enhanced thermoelectric properties of their nanocomposites. *Adv. Mater.* **2013**, *25*, 1425–1429. [CrossRef] [PubMed]

23. Mehta, R.J.; Zhang, Y.; Karthik, C.; Singh, B.; Siegel, R.W.; Borca-Tasciuc, T.; Ramanath, G. A new class of doped nanobulk high-figure-of-merit thermoelectrics by scalable bottom-up assembly. *Nat. Mater.* **2012**, *11*, 233–240. [CrossRef] [PubMed]

24. Gheiratmand, T.; Hosseini, H.R.M.; Davami, P.; Ostadhossein, F.; Song, M.; Gjoka, M. On the effect of cooling rate during melt spinning of finemet ribbons. *Nanoscale* **2013**, *5*, 7520–7527. [CrossRef] [PubMed]

25. Tkatch, V.I.; Limanovskii, A.I.; Denisenko, S.N.; Rassolov, S.G. The effect of the melt-spinning processing parameters on the rate of cooling. *Mater. Sci. Eng. A* **2002**, *323*, 91–96. [CrossRef]

26. Jiang, H.; Moon, K.-S.; Dong, H.; Hua, F.; Wong, C.P. Size-dependent melting properties of tin nanoparticles. *Chem. Phys. Lett.* **2006**, *429*, 492–496. [CrossRef]

27. Donovan, E.; Spaepen, F.; Turnbull, D.; Poate, J.; Jacobson, D. Heat of crystallization and melting point of amorphous silicon. *Appl. Phys. Lett.* **1983**, *42*, 698–700. [CrossRef]

28. Xu, Z.J.; Hu, L.P.; Ying, P.J.; Zhao, X.B.; Zhu, T.J. Enhanced thermoelectric and mechanical properties of zone melted p-type $(Bi,Sb)_2Te_3$ thermoelectric materials by hot deformation. *Acta Mater.* **2015**, *84*, 385–392. [CrossRef]

29. Xie, W.; Tang, X.; Yan, Y.; Zhang, Q.; Tritt, T.M. Unique nanostructures and enhanced thermoelectric performance of melt-spun BiSbTe alloys. *Appl. Phys. Lett.* **2009**, *94*, 102111. [CrossRef]

© 2017 by the authors. Licensee MDPI, Basel, Switzerland. This article is an open access article distributed under the terms and conditions of the Creative Commons Attribution (CC BY) license (http://creativecommons.org/licenses/by/4.0/).

crystals

MDPI

Article

Structural and Electrical Properties Characterization of $Sb_{1.52}Bi_{0.48}Te_{3.0}$ Melt-Spun Ribbons

Viktoriia Ohorodniichuk [1], Anne Dauscher [1], Elsa Branco Lopes [2], Sylvie Migot [1], Christophe Candolfi [1] and Bertrand Lenoir [1,*]

[1] Institut Jean Lamour, UMR 7198 CNRS, Université de Lorraine, Parc de Saurupt, CS 50840, 54011 Nancy, France; viktoriia.ohorodniichuk@univ-lorraine.fr (V.O.); anne.dauscher@univ-lorraine.fr (A.D.); sylvie.migot@univ-lorraine.fr (S.M.); christophe.candolfi@univ-lorraine.fr (C.C.)

[2] C2TN, Instituto Superior Técnico, Universidade de Lisboa, Estrada Nacional 10, 2695-066 Bobadela LRS, Portugal; eblopes@itn.pt

*   Correspondence: bertrand.lenoir@univ-lorraine.fr; Tel.: +33-383-584163

Academic Editor: George S. Nolas

Received: 28 April 2017; Accepted: 7 June 2017; Published: 13 June 2017

Abstract: Melt-spinning (MS) has been reported as a promising tool to tailor the microstructure of bulk thermoelectric materials leading to enhanced thermoelectric performances. Here, we report on a detailed characterization of *p*-type $Bi_{0.48}Sb_{1.52}Te_3$ ribbons produced by melt-spinning. The microstructure of the melt-spun ribbons has been studied by means of X-ray diffraction, scanning and transmission electron microscopy (TEM). The analyses indicate that the ribbons are highly-textured with a very good chemical homogeneity. TEM reveals clear differences in the microstructure at large and short-range scales between the surface that was in contact with the copper wheel and the free surface. These analyses further evidence the absence of amorphous regions in the melt-spun ribbons and the precipitation of elemental Te at the grain boundaries. Low-temperature electrical resistivity and thermopower measurements (20–300 K) carried out on several randomly-selected ribbons confirm the excellent reproducibility of the MS process. However, the comparison of the transport properties of the ribbons with those of bulk polycrystalline samples of the same initial composition shows that MS leads to a more pronounced metallic character. This difference is likely tied to changes in deviations from stoichiometry due to the out-of-equilibrium conditions imposed by MS.

Keywords: melt-spinning; microstructure; X-ray diffraction; transmission electron microscopy; electrical properties

---

1. Introduction

Thermoelectric materials provide a versatile, environmentally-friendly way for generating electric power from waste heat or for Peltier cooling [1,2]. Despite the fact that this technology has been successfully used for decades to power deep-space probes and rovers, its applications remain limited to niche technologies for which the robustness of thermoelectric modules outweighs their low conversion efficiency [1,2]. A more widespread use of thermoelectric materials is therefore tied to the identification of novel families of materials exhibiting a high dimensionless thermoelectric figure of merit:

$$ZT = \alpha^2 T / \rho\kappa \tag{1}$$

where α is the thermopower, ρ is the electrical resistivity, κ is total the thermal conductivity and T is the absolute temperature; or to the optimization of the thermoelectric properties of state-of-the-art materials [1–3].

Most of the recently-discovered families of thermoelectric materials reach their maximum ZT values at high temperatures, typically between 700 and 1200 K [4–12]. For thermoelectric applications near room temperature, solid solutions of bismuth telluride Bi_2Te_3 with the isomorphous compounds Sb_2Te_3 and Bi_2Se_3 are still nowadays the best materials with ZT values around unity in both p- and n-type compounds [2]. The $A^V_2B^{VI}_3$ (A = Bi or Sb and B = Te or Se) compounds and their solid solutions crystallize in the $R\bar{3}m$ space group and are narrow-band-gap semiconductors with topologically-protected gapless surface states [2,13–16]. Their crystal structure is composed of repeated planes of five-atomic layer lamellas perpendicular to the trigonal axis separated by a van der Waals gap. This layered structure inevitably results in anisotropic transport properties in both single-crystalline and polycrystalline specimens [2].

Several synthesis techniques were used in the past to prepare these compounds either in single or bulk polycrystalline form [17–25]. Polycrystals are traditionally fabricated using the zone melting method, powder metallurgy techniques or mechanical alloying, followed by a consolidation step [26–31]. The melt-spinning (MS) method, based on rapid solidification of the melt that allows attaining cooling rates as high as 10^4–10^7 K s^{-1}, has been used to produce these materials by Soviet Union and Russian research groups [32–35]. Starting from raw materials, MS produces ribbons, flakes or foils, which are in an out-of-equilibrium state due to the high quenching rate leading to particular microstructures and physical properties. This technique has also been the subject of studies from a mathematical point of view [36]. Recently, this technique combined with subsequent hot pressing was employed successfully by several groups to achieve high thermoelectric performance with peak ZT values of ~1.5 around 300 K in p-type $Bi_xSb_{2-x}Te_3$ for $0.48 \leq x \leq 0.52$ [37–43]. Because melt-spun ribbons are subsequently consolidated to obtain bulk dense specimens, a detailed investigation of their microstructure-properties relationships is essential to better understand the influence of the MS process on the thermoelectric properties of Bi_2Te_3-based materials. In this context, Koukharenko et al. investigated the microstructure and transport properties of ribbons obtained from Bi_2Te_3 and from the $Bi_{2-x}Sb_xTe$ and $Bi_{2-x}Sb_xTe_2$ systems [44–49]. However, melt-spun ribbons in the ternary p-type $Bi_xSb_{2-x}Te_3$ solid solution ($0.48 \leq x \leq 0.52$) have received much less attention even though the best thermoelectric performances are achieved for these particular compositions [2].

Here, we report on a detailed investigation of the microstructure and chemical homogeneity of $Bi_{0.48}Sb_{1.52}Te_3$ ribbons produced by melt-spinning along with low-temperature electrical resistivity and thermopower measurements (20–300 K) performed on several randomly-selected ribbons. We find that the $Bi_{0.48}Sb_{1.52}Te_3$ melt-spun ribbons exhibit a complex, highly-textured microstructure with a very good chemical homogeneity. The transport properties of different ribbons do not show any significant deviation to within experimental uncertainty, confirming the high reproducibility achieved with the MS technique. Yet, our results evidence that the melt-spun ribbons show a more pronounced metallic character with respect to bulk polycrystalline samples of the same initial composition, which highlights the extreme sensitivity of the $Bi_xSb_{2-x}Te_3$ compounds to the synthetic process used. We attribute this difference to modifications in the deviations from stoichiometry as a result of the strong out-of-equilibrium conditions achieved in the MS process.

2. Experimental Details

2.1. Synthesis

The production of ribbons by the melt-spinning technique was realized by a two-step process. As a first step, an ingot of composition $Bi_{0.48}Sb_{1.52}Te_3$ was prepared from stoichiometric amounts of high-purity elements in the form of granules (Bi, Sb and Te, 5N+, 99.999%). They were loaded in quartz tubes (previously cleaned in acids and evacuated) and maintained under secondary vacuum for 3 h. The ampoule was then sealed under a reducing atmosphere composed of a mixture of H_2 and He (5/95%). The tube was kept at 983 K during 5 h in a vertical oscillating furnace followed by a quenching in a room-temperature water bath. A part of the resulting ingot was crushed into fine powders using an agate mortar. The powder was consolidated by spark plasma sintering (SPS) at 773 K under 30 MPa for 5 min in graphite dies. The relative density of the samples, determined from weight and sample dimensions, was above 95% of the theoretical density.

2.2. Melt-Spinning Process

The second part of the ingot was used for the MS process carried out with a melt-spinner (Edmond Bühler) equipped with a copper wheel of ~20 cm in diameter. Approximately 10 g of the solid ingot were placed in quartz tubes with a *V*-shaped end and a nozzle diameter of 1 mm. The ingot was heated up to 893 K under argon atmosphere (around 0.6 bar), the temperature inside the tube being continuously checked by a MAURER digital infrared pyrometer. The melt was ejected on the copper wheel rotating at $U = 35$ m s^{-1} (linear speed) using an overpressure of 0.8 bar of argon. The melt was instantaneously cooled on the water-cooled wheel forming "ribbons", "foils" or "flakes" with typical dimensions 3–5 mm in length, 0.5–3 mm in width and 8 ± 1 μm in thickness [50]. This last value was obtained from cross-section observations with scanning electron microscopy (SEM) of more than 20 ribbons that were randomly selected.

The MS process has been visualized using an ultra-high-speed video system (Photron SA5) with a frame rate of 12,000 fps. This system enables estimating the average residence time:

$$\tau = L/U \tag{2}$$

Defined as the time it takes for a point on the wheel surface to rotate through the length L of the puddle, estimated to be ~1 mm in our case. This yields an average residence time of roughly 30 μs. As shown by Huang et al. [51], τ is correlated to the ribbon thickness R, such that $R \approx \tau^{1/2} \times 10^{-3}$ m s$^{-1/2}$. Taking into account the above-mentioned value of τ, the thickness of the ribbons should be ~5 μm, that is close to the experimental values obtained by SEM.

The cooling rate K achieved in our experiments was estimated following the relation used by Fedotov et al. [52] for experiments performed under static conditions:

$$K = a\theta/C_p dR \tag{3}$$

where a is the heat convection coefficient, θ is the excessive temperature of the melt, C_p is the specific heat of the melt (~0.19 J g^{-1} K^{-1}) and d is the density of $Bi_{0.48}Sb_{1.52}Te_3$ (6.88 g cm^{-3}). The heat convection coefficient for a polished copper surface is estimated to be in the range of 1–2×10^5 W m^{-2} K^{-1} [53]. In the literature, the excessive temperature of the melt is defined as the difference between the temperature of the melt and the temperature of the wheel considered to be equal to room temperature [46]. For ribbons with a thicknesses close to 8 μm, the cooling rate is then estimated to be of the order of 10^6 K s^{-1}.

2.3. Structural and Chemical Characterizations

X-ray diffraction (XRD) analyses of the ribbons were carried out with a Bruker D8 Advance diffractometer in Bragg–Brentano geometry using Cu Kα_1 radiation ($\lambda = 1.54056$ Å). Besides analyzes on ground ribbons, the two surfaces of the ribbons, i.e., the surface in contact with the wheel and the free surface, were checked in order to unveil possible different textures. We note that the X-ray penetration depth is small enough in these materials so that only a part of the ribbon's volume underneath the surface is probed. This hypothesis is confirmed by the estimation of the penetration depth δ for normal incidence that specifies the path length for which the intensity drops to $1/e$ of its initial value:

$$\delta_{1/e} = (d\Sigma_i(\mu_m)_i g_i / G)^{-1} \tag{4}$$

where g_i is the atomic mass of the element in the compound, G is the mass of the molecular unit and μ_m is the mass absorption coefficient of the element [54]. For Cu Kα radiation, the μ_m values are 25.9, 26.7

and 24.4 m^2 kg^{-1} for Sb, Te and Bi, respectively [55]. For the composition Bi$_{0.48}$Sb$_{1.52}$Te$_3$, this relation yields an estimated penetration depth δ of 5.6 µm, that is less than the average ribbon thickness.

The microstructure of the ribbons was checked by using two different field emission gun (FEG) scanning electron microscopes (SEM-FEG XL30 and Quanta 650 FEG both from FEI). The ribbons were observed on both faces and cross-sections. The thickness and width of all ribbons used for electrical measurements were estimated by cross-section and top views. The composition and chemical homogeneity at the micrometric scale were determined by energy dispersive X-ray spectrometry (EDS, Bruker, Wissembourg, France) mounted on the Quanta microscope.

To perform transmission electron microscopy (TEM) studies, thin slices of the melt-spun ribbons (surfaces and cross-sections) were prepared by FEI company by the dual focused ion beam (FIB)-scanning electron microscope (SEM) system using the "in situ" lift-out technique. Transmission electron microscopy (TEM) investigations were performed on a Philips CM-200 microscope (Eindhoven, The Netherlands) operating at 200 kV to check the quality of the thin slices. TEM, high-resolution TEM (HRTEM) and scanning TEM (STEM) associated with high-angle annular dark-field (HAADF) were also performed on a JEOL ARM 200F-Cold FEG TEM/STEM microscope (Tokyo, Japan) running at 200 keV and equipped with a GIF Quantum ER.

2.4. Transport Measurements

The dense SPS Bi$_{0.48}$Sb$_{1.52}$Te$_3$ pellet, used as a reference, was cut both parallel and perpendicular to the pressing direction with a diamond wire-saw into bar-shaped samples of typical dimensions $2.5 \times 3.0 \times 8.0$ mm^3. Electrical resistivity and thermopower were simultaneously measured between 5 and 300 K in the continuous mode with the thermal transport option (TTO) of a physical property measurement system (PPMS, Quantum Design, San Diego, CA, USA). The electrical and thermal contacts were made by brazing four copper bars with a low melting point braze. The experimental uncertainty on resistivity and thermopower is estimated to be 5%.

Electrical resistivity and thermopower measurements were carried out in the ~20–300 K temperature range on several melt-spun ribbons (with a plate-like shape of typical dimensions ~400 µm in length and with a section of 350×8 µm^2; see below) randomly chosen using a dedicated cell attached to the cold stage of a closed-cycle refrigerator, the details of which are provided elsewhere [56]. Both current and thermal gradient were applied along the length of the ribbons. The electrical resistivity was measured by a four-probe AC method, using an SRS Model SR83 Lock-in Amplifier with a low-frequency current of 5 mA (77 Hz) applied to the sample. The thermopower was measured by a slow AC technique (ca. 10^{-2} Hz), the voltage across the sample and gold leads being measured with a Keythley 181 nanovoltmeter. The oscillating thermal gradient was kept below 1 K and was measured by a Au-0.005 at % Fe versus chromel thermocouple. The absolute thermopower of the sample was obtained after correction for the absolute thermopower of the gold leads (99.99% pure gold) by using the data of Huebener [57]. The experimental uncertainties on the electrical resistivity and thermopower are estimated to be 7% and 5%, respectively.

The Hall resistivity ρ_H was determined on bulk SPS samples and ribbons from measurements of the transverse electrical resistivity ρ_{xy} under magnetic fields $\mu_0 H$ ranging between -1 and $+1$ T using the AC transport option of the PPMS at room temperature. The data were corrected for slight misalignment of the contacts by applying the formula:

$$\rho_H = [\rho_{xy}(\mu_0 H) - \rho_{xy}(-\mu_0 H)]/2 \tag{5}$$

The Hall coefficient R_H was determined from the slope of the $\rho_H(\mu_0 H)$ data in the limit $\mu_0 H \to 0$. The Hall carrier concentration p and mobility μ_H were estimated within a single-band model with a Hall factor r_H equal to 1 that yields the relations:

$$p = r_H/R_H e = 1/R_H e \tag{6}$$

and:

$$\mu_H = R_H/\rho \tag{7}$$

3. Results and Discussion

3.1. X-ray Diffraction and Scanning Electron Microscopy

Figure 1 shows the XRD patterns collected on ground ribbons, as well as on the surface in contact with the copper wheel and the free surface. The patterns show that the ribbons are well crystallized regardless of the sample considered. No significant peak broadening is observed suggesting that the grain size is above the nanoscale range. All of the reflections for the ground ribbons and free surface can be indexed with the standard polycrystalline pattern of $Bi_{0.48}Sb_{1.52}Te_3$ indicating the absence of impurity phases. If the patterns of those samples are quite similar, that of the contact surface clearly exhibits a significant degree of texturing along the (110), (015) and (125) planes, indicating an orientation effect during the material's solidification. The trigonal axis of the crystallites orientated along these three directions forms an angle of 90°, 58°45′ and 77°4′ respectively, with the normal of the free surface as shown in Figure 2. As the XRD pattern of the free surface is similar to that of ground ribbons, the level of texturing should be limited to a region close to the contact surface. Our results are however quite different from those obtained by Koukarenko et al. [46], who studied in detail the texture formation in Bi_2Te_3 ribbons. Their investigation revealed a well-defined (025) texture independent of the quenching temperature, the ribbon thickness and the heat treatment. This (025) texture was proposed to be correlated to the nature of the Bi-Te covalent bond in this plane. Further investigations on the $Bi_{2-x}Sb_xTe_3$ ($0 \leq x \leq 1$) system by the same authors showed that substituting Sb for Bi tends to lessen the (025) texture and favors the appearance of the (110) texture [48]. The difference between the textures observed in [46] and in our case could be linked to different cooling rates since the thickness of the ribbons obtained in their studies was significantly higher (between 20 and 35 μm).

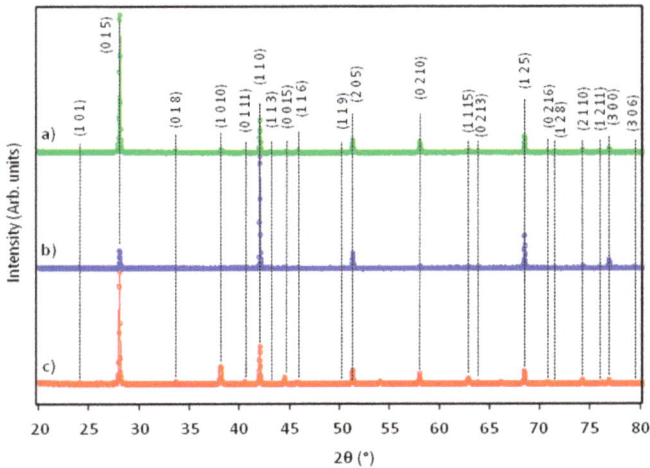

Figure 1. XRD patterns of the free surface (**a**), contact surface (**b**) and ground ribbons (**c**) of melt-spun $Bi_{0.48}Sb_{1.52}Te_3$.

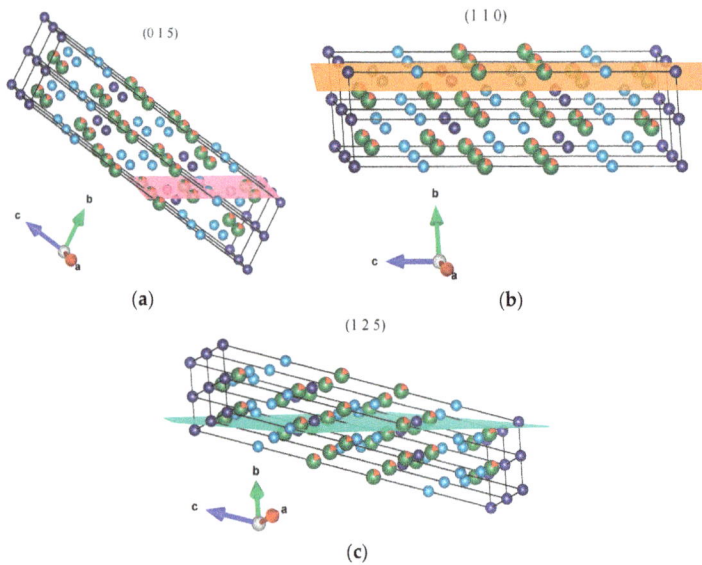

Figure 2. Scheme of the most pronounced orientations depending on the surface of the melt-spun ribbons probed: (**a**) free surface (015) planes; (**b**) wheel contact surface (110) planes; (**c**) wheel contact surface (125) planes. The Bi, Sb, Te_1 and Te_2 atoms are shown in red, green, dark blue and light blue, respectively. The two-colored atoms correspond to the mixed occupation of Bi and Sb with the ratios of 24% and 76%, respectively.

The obtained microstructures observed by SEM are similar to those observed in prior studies on similar or close compositions (Figure 3) [37–39,43,58]. Typically, top views of the contact surface do not demonstrate any specific microstructural details except those related to the wheel roughness (Figure 3a), while top views of the free surface exhibit what Xie et al. [37–39,43] called a dendritic-like microstructure (or a needle network microstructure, as termed by Koukarenko et al. [46,48]) of 0.1–0.5 μm in width (Figure 3b). Cross-section views of the ribbons, shown in Figure 3c, indicate a close-packed microstructure of about some hundreds of nanometers in thickness in the region close to the contact surface, which is composed of a mixture of small-sized particles up to one micrometer in size and possibly of nanosized particles as expected from the MS process. This structure is followed by a columnar growth of the basal planes giving rise to a dendritic or a needle-like appearance of the free surface. According to the XRD results, preferential growth occurs in the thin close-packed microstructure, while an erratic growth takes place in the columnar structure.

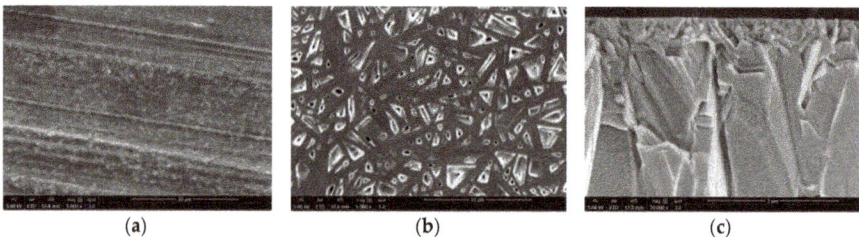

Figure 3. SEM images of a top view of the contact surface (**a**), of a top view of the free surface (**b**) and a cross-section (**c**) of part of the ribbon highlighting the close-packed microstructure at the contact surface of a $Bi_{0.48}Sb_{1.52}Te_3$ ribbon.

The spatial distribution of the elements has been assessed by elemental X-ray mapping on the free and contact surfaces, as well as on a cross-section of the ribbons (Figure 4). At the scale probed by these experiments, all of the elements appear homogeneously distributed within the ribbon. The atomic composition determined by EDXS $Bi_{0.4}Sb_{1.6}Te_{3.0}$ is very close to the expected composition, given the experimental uncertainty that stems from the strong overlap of the La lines of Te and Sb.

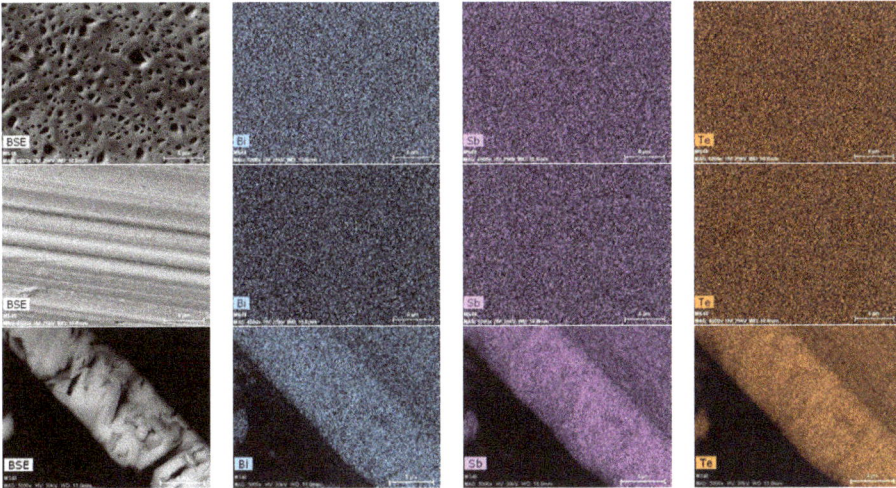

Figure 4. Backscattered electron (BSE) images and corresponding X-ray elemental mappings of: (**top line**) the free surface; (**middle line**) the contact surface; and (**bottom line**) the cross-section of a $Bi_{0.48}Sb_{1.52}Te_3$ ribbon.

3.2. Transmission Electron Microscopy

TEM and HRTEM studies have been carried out to gain relevant insights into the nanostructure of the ribbons. We address in particular the issues concerning the formation of amorphous zones during the MS process, the presence of which has not been systematically observed in prior studies, and the possible precipitation of elemental Te as observed in single-crystals grown in out-of-equilibrium conditions. A top view obtained by TEM of one of the thin cross-section slices produced by FIB is shown in Figure 5a. Similar features to SEM observations can be seen, that is a close-packed microstructure close to the contact surface of less than 1 μm in thickness followed by a columnar growth. The thickness of the columnar grains is less than 1 μm. The surface close to the wheel is composed of grains of about 100 nm (Figure 5b). In this analyzed section, neither an amorphous, nor a nano-sized zone could be observed. This observation contrasts with the results obtained by Xie et al. [39,42], who found the presence of an amorphous layer of about 500 nm in thickness for an overall thickness of about 3 mm (assuming that the authors showed the entire cross-section). The presence of nanoparticles embedded in an amorphous matrix (thickness of about 1 mm as shown in [39]) could not be observed either. The reasons for these discrepancies remain so far unknown.

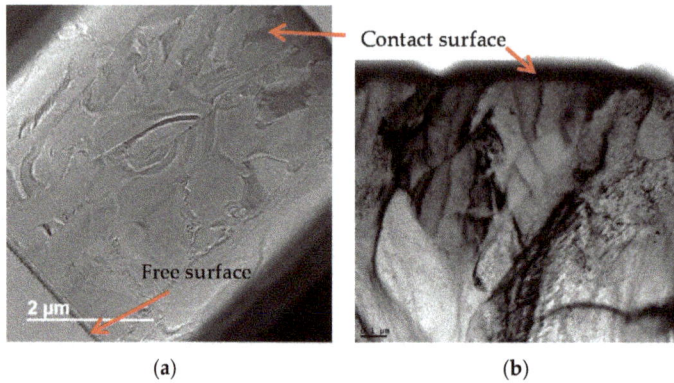

(a) **(b)**

Figure 5. TEM images of (**a**) a thin slice of a cross-section of a $Bi_{0.48}Sb_{1.52}Te_3$ ribbon (overall thickness = 8.1 mm). The top is covered with an amorphous carbon layer and the bottom with a copper layer, due to the preparation process of the thin slice. (**b**) Contact surface made of grains of about 100 nm. The top black layer is amorphous carbon.

Figure 6 displays two HRTEM images collected in the middle zone of a ribbon. The first image was taken at the interface between two grains, while the second image was taken in the center of a grain. Based on the fast Fourier transform (FFT), the grains are well crystallized, and the average inter-fringe distances d taken along the lines in the main and perpendicular directions (0.319 and 0.213 nm, respectively) are in good agreement with the inter-planar distances of the (015) (0.317 nm) and (110) (0.215 nm) planes [59]. These measurements further confirm the orientations found in our XRD analyses. Interestingly, some nano-sized particles of elemental Te could be also observed at grain boundaries as illustrated in Figure 7.

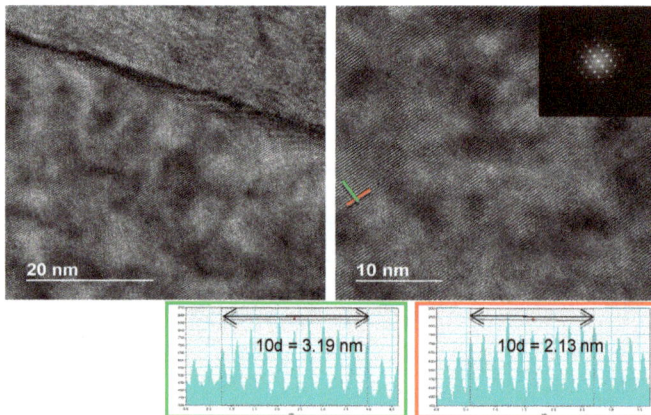

Figure 6. HRTEM images of an interface between two grains (**left**) and of a well-crystallized grain (**right**) taken in the center of a $Bi_{0.48}Sb_{1.52}Te_3$ melt-spun ribbon. Inter-planar fringe distances have been evaluated through the line profiles taken on the rows highlighted in the right image. The distances $10d$ (line profiles) are given in nm.

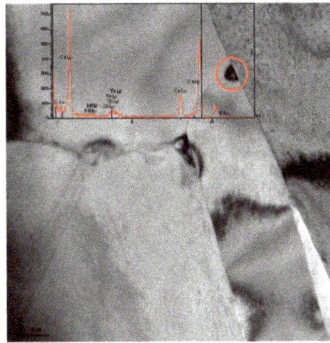

Figure 7. TEM image emphasizing the presence of nano-sized Te precipitates in the ribbons.

Figure 8 shows a grain boundary between four adjacent grains in the dendritic-like zone obtained from an SEM-FIB thin slice cut parallel to the free surface. The triangular-shaped feature of about 70 nm on the side and located in between larger grains is typical of growth with the trigonal axis perpendicular to the surface evidenced by the hexagonal disposition of the spots obtained on the FFT of this zone. Some of these triangular-shaped structures that exhibit clean surfaces with the surrounding crystals can be observed all over the thin slice. The tops of the two columnar grains surrounding the triangular crystal are single crystalline. Both grains display an interplanar distance of 0.117 nm that corresponds to the (205) planes. Such sub-micron-sized crystalline domains were systematically observed in prior investigations on MS ribbons of $Bi_xSb_{2-x}Te_3$ [37–39,42,60].

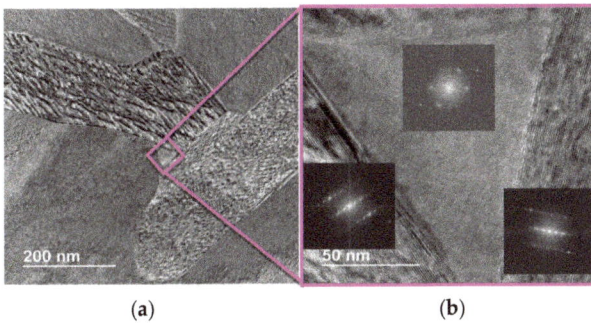

(a)   (b)

Figure 8. TEM images of a grain boundary between four adjacent grains (a) and magnification of the highlighted square (b) taken in the dendritic zone of a $Bi_{0.48}Sb_{1.52}Te_3$ ribbon.

TEM images also reveal lattice stripes with a larger estimated width of about 10 Å (Figure 9). A magnification of this zone shows that these stripes are formed by a series of five bright-doted rows separated by a more dark-spotted row. This result is consistent with the five-layer lamellae structure of the ideal crystal structure of Sb-Bi tellurides. The above-mentioned width is also in very good agreement with the height of the -Te-Bi-Te-Bi-Te- quintet (12 Å according to [61]). These features are not inherent to the MS process we employed and have also been reported by Lan et al. [62] and Li et al. [63] in *p*-type $Bi_xSb_{2-x}Te_3$ samples prepared by two synthetic routes consisting of mechanical alloying-SPS and mechanical alloying-hot pressing, respectively. In these two studies, however, the appearance of these stripes was slightly different and was described as a series of two rows of extra-bright dots separated by four weaker bright-dot rows forming the five-layer lamellae.

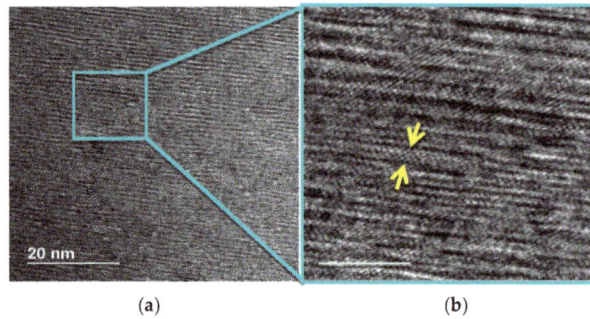

Figure 9. HRTEM images showing lattice stripes of about 1 nm in width (**a**) and magnification of the highlighted square (**b**). The arrows show the quintet organization of a $Bi_{0.48}Sb_{1.52}Te_3$ grain.

In addition, we observed a needle-like structure containing nano-grains of about 5–10 nm in size in another thin slice (Figure 10). In contrast to the larger grains, these small grains do not seem to be closely packed. The needle was located in the middle of the cross-section of a ribbon. It is however difficult to determine in this case whether the diffuse halo rings observed in the FFT are due to the MS process or to the small amount of amorphized material produced by the impact of high-energy Ga ions during the FIB sample thinning. We note that we did not observe nanostructures appearing as dense striations with spacing of the order of 10 nm reported in Bi_2Te_3 by Jacquot et al. [60] and Lan et al. [62] and described in the prior study of Peranio and Eibl [64]. Although these structures could be present as well in our samples, these features may also arise from the preparation of the thin slices by ion-milling with Ar^+ ions as underlined by Homer and Medlin [65].

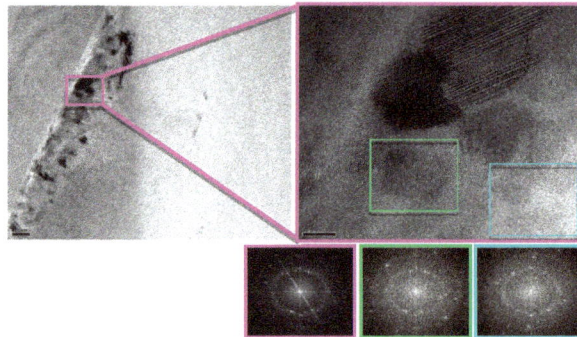

Figure 10. TEM (**left**) and HRTEM (**right**) images of a needle-like structure containing nano-grains of about 5–10 nm in size.

The chemical composition of the ribbons was further analyzed by EDXS using the STEM mode along five profiles on two ribbons as depicted in Figure 11. The compositions seem independent of the region probed, the data taken along a columnar grain (green and sky blue lines) and in the dendritic zone (dark blue and red lines) being roughly similar. X-ray elemental mappings were performed in several zones of one ribbon, that is in the dendritic zone, at the interface of the amorphous-crystallized grains and within a hexagonal grain (Figure 12). The elemental distribution shows that the elements are evenly distributed in the dendrites, the columnar structures and the amorphous layer, without any particular compositional segregation.

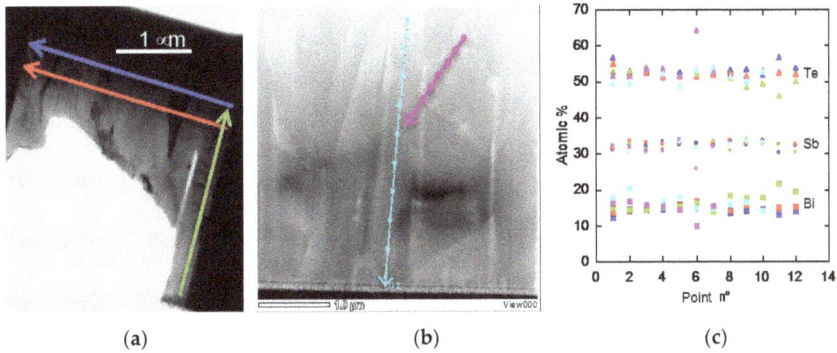

(a) (b) (c)

Figure 11. EDXS results for as-grown $Sb_{1.52}Bi_{0.48}Te_3$ ribbons measured along the colored profiles. The *x*-axis corresponds to the number of the points analyzed. The results are shown in the graph using the same color code for two different samples analyzed.

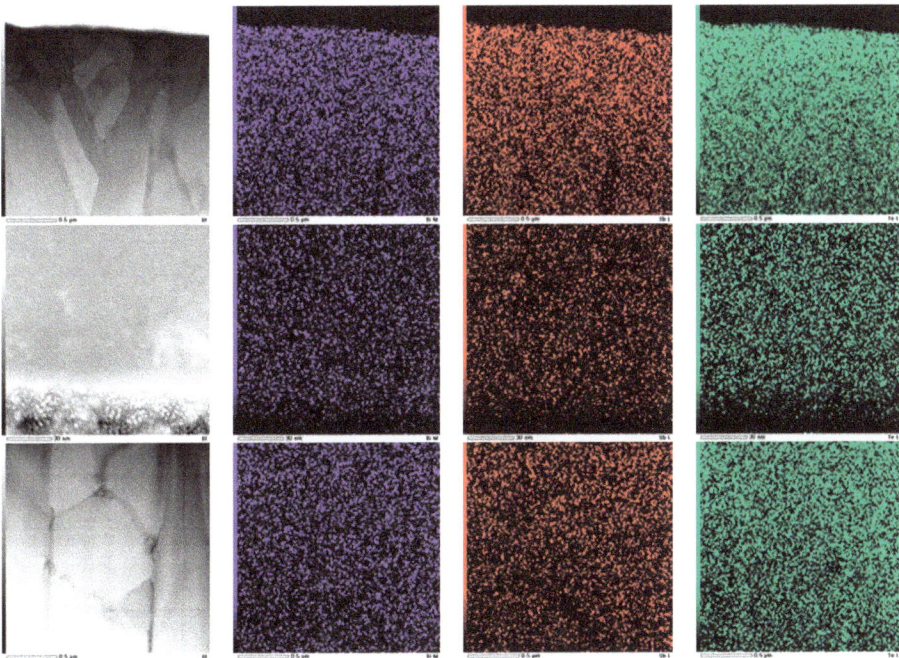

Figure 12. EDXS elemental mapping images of an as-grown $Bi_{0.48}Sb_{1.52}Te_3$ ribbon taken at three different places. HAADF-STEM image along with the corresponding Bi (blue), Sb (red) and Te (green) elemental maps.

3.3. Transport Properties

Figure 13a shows the temperature dependence of the electrical resistivity ρ of three randomly-selected ribbons along with the data collected on the reference $Bi_{0.48}Sb_{1.52}Te_3$ bulk polycrystalline sample measured parallel and perpendicular to the pressing direction. The ρ values of the bulk sample are constant below 20 K and increase above this temperature with increasing temperature. Above 200 K, ρ roughly follows a $T^{1.5}$ law. Taking into account the magnitude of ρ

(from 1 up to 12 $\mu\Omega$ m), this behavior is typical of heavily-doped semiconductors as expected for the $Bi_{0.48}Sb_{1.52}Te_3$ composition. Consistent with the results obtained on single crystals [23,24], these measurements show a significant difference in the data measured perpendicular and parallel to the pressing direction, the latter of which is higher. Further, the anisotropy ratio defined as:

$$\gamma = \rho_{par}/\rho_{perp} \tag{8}$$

is not constant over the whole temperature range, but increases with temperature to reach 1.8 at 300 K. This result indicates that the SPS process induces a preferred orientation, which is parallel to the trigonal axis, along the SPS pressing direction. The $\rho(T)$ data measured on the ribbons show the same general trend with respect to the reference sample. The ρ values fall in between those of the bulk samples with room-temperature values approaching 10 mΩ m. These measurements tend to indicate that the typical microstructure of the ribbons does not significantly affect the electrical transport. Of note is the fact that the three randomly-selected ribbons show similar values to within experimental uncertainty, which evidences that MS is a robust process for producing chemically-homogeneous ribbons with nearly-identical electrical properties.

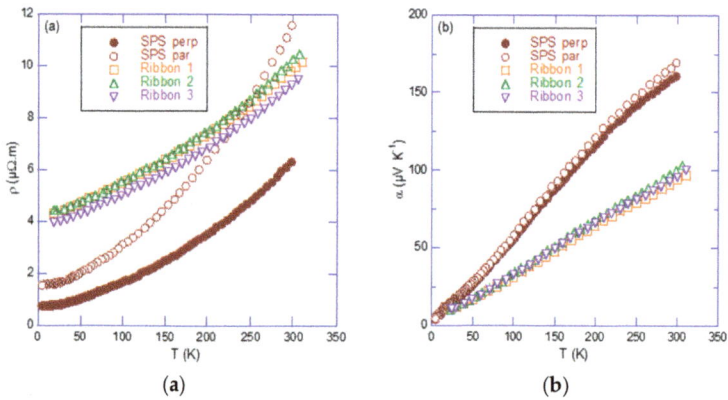

Figure 13. Temperature dependence of the electrical resistivity ρ (**a**) and thermopower α (**b**) for the bulk sample consolidated by spark plasma sintering (SPS) probed along and perpendicular to the pressing direction and for three randomly-selected ribbons.

The thermopower α of the $Bi_{0.48}Sb_{1.52}Te_3$ bulk sample and ribbons is shown in Figure 13b as a function of temperature. All samples exhibit positive α values, which vary linearly below 40 K and logarithmically above 200 K. In agreement with the ρ data, this behavior is consistent with those of heavily-doped semiconductors. Our results obtained on the SPS sample are in line with the well-known fact that α does not depend on the orientation in the extrinsic or one-carrier regime in (Sb_2Te_3)-(Bi_2Te_3) solid solutions [2]. No differences in the α values for the three ribbons are visible, confirming their very similar physical properties. Yet, the α values of the ribbons are lower than those of the bulk specimen in the whole temperature range. This behavior is likely related to a change of either the carrier concentration and/or the scattering parameter due to the MS process. Measurement of the Hall coefficient R_H reveals that the apparent Hall concentration p_H differs significantly at 300 K for the bulk sample and the ribbons (Table 1). The hole concentration measured in the ribbons is almost two-times higher than in the bulk, indicative of the more pronounced metallic character of the ribbons. Thus, the very good agreement between the electrical resistivity observed in the SPS samples and in the ribbons is only fortuitous and is a direct consequence of the increased hole concentration that compensates the decrease in the Hall mobility μ_H (Table 1). The degradation of μ_H is likely linked to the numerous interfaces at the microscale length present along the current direction, as evidenced by microscopic analyses, which efficiently scatter holes. Note that a more metallic state has also been observed at 300 K in $Bi_xSb_{2-x}Te_3$ ribbons for $x = 0.40$ [48]. In these prior investigations, the melt-spinning technique was shown to produce homogeneous ribbons with identical electrical properties although the intrinsic properties of the carriers (concentration and mobility) are different from those of the bulk SPS samples. The excellent reproducibility of the transport properties in $Bi_{0.48}Sb_{1.52}Te_3$ ribbons observed herein thus confirms the conclusions drawn for analogous compounds [43–49,60].

Table 1. Room-temperature values of the Hall coefficient R_H, electrical resistivity ρ, Hall concentration p_H and Hall mobility μ_H of the SPS bulk sample (perpendicular to the pressing direction) and measured on one ribbon.

	R_H (cm³ C⁻¹)	ρ (μΩ m)	p_H (10^{19} cm⁻³)	μ_H (cm² V⁻¹ s⁻¹)
SPS	0.17	6.4	3.7	266
Ribbon	0.07	10	9.1	70

Finally, it should be kept in mind that the electronic properties of mixed crystals based on Sb_2Te_3-Bi_2Te_3 are governed by native defects, the influence of which outweighs the role of differences in the microstructure of the samples. In Sb-rich Sb_2Te_3-Bi_2Te_3 compositions, antistructure defects are the main type of defects with Sb (and partially Bi) atoms replacing Te atoms, which are usually denoted Sb_{Te} (or Bi_{Te}). The Sb_{Te} defects are electrically active and behave as single acceptors giving rise to *p*-type electrical conduction [66,67]. The presence of native defects prevents the solid solution from being stoichiometric, that is the higher the defect concentration, the higher the deviation from stoichiometry. Controlling these deviations, and hence the electrical properties, is the main challenge to overcome for optimizing the thermoelectric properties of these compounds. In this context, the precise knowledge of the phase diagram can be a powerful tool to properly control the carrier concentration. The solidus line, characterizing the maximum deviation from stoichiometry, was investigated in the past in the Te and Sb rich-side of the $Sb_{2-x}Bi_xTe_3$ ($x = 0.0, 0.4$ and 0.5) solid solution at the thermodynamic equilibrium [2,23]. However, for non-equilibrium processes, the phase diagram can be severely affected as demonstrated in prior studies [17,68]. These works showed that in fast-cooling processes, Te exhibits a retrograde solubility and thus tends to precipitate out of the main phase, in agreement with our TEM observations. Since a lower Te content in the matrix contributes to further enhancing the deviations from stoichiometry, it results in an increased hole concentration due to the triple-acceptor nature of Te vacancies [67]. The slight Te precipitation is thus likely at the origin of the increased hole concentration we observe in melt-spun ribbons.

4. Conclusions

p-type $Bi_{0.48}Sb_{1.52}Te_3$ ribbons have been synthesized successfully via melt-spinning. Our detailed structural and chemical characterizations have confirmed that this technique enables achieving excellent chemical homogeneity and reproducibility. TEM studies carried out on ribbons revealed the very different microstructure exhibited by the free surface and the surface in direct contact with the copper wheel. The very high cooling rates achieved with this technique have a sizeable influence on the transport properties of the ribbons that exhibit a more metallic nature compared to bulk samples of the same initial composition prepared from a conventional synthesis route. Variations in deviations from stoichiometry, which are known to play a prominent role in this family of compounds, likely explain this difference. These results provide a good basis to better understand the variations in the transport properties of *p*-type $Bi_xSb_{2-x}Te_3$ compounds at various steps of the synthetic process used to produce bulk samples from melt-spun ribbons. Extending these investigations to other compositions would be of interest to determine whether melt-spinning leads to enhanced thermoelectric performances in Bi_2Te_3-based solid solutions.

Acknowledgments: This work was supported by Electricité de France R&D through the CIFRE (Conventions Industrielles de Formation par la Recherche), convention No. 2011/1329. The authors thank Pascal Dalicieux, Philippe Baranek and Laurent Legras from Electricité de France. They are also grateful to Jaafar Ghanbaja for his support in HRTEM investigations.

Author Contributions: B.L. conceived of and designed the experiments. V.O. performed the experiments, prepared all of the samples for characterizations and transport properties measurements and analyzed the data. S.M. and A.D. performed the TEM investigations. A.D. carried out SEM experiments. E.B.L. performed the measurements on the ribbons. V.O., B.L. and C.C. wrote the paper.

Conflicts of Interest: The authors declare no conflict of interest. The founding sponsors had no role in the design of the study; in the collection, analyses or interpretation of data; in the writing of the manuscript; nor in the decision to publish the results.

1. Goldsmid, H.J. *Thermoelectric Refrigeration*; The International Cryogenics Monograph Series; Plenum Press: New York, NY, USA, 1964.
2. Rowe, D.M. *Thermoelectrics and Its Energy Harvesting*; CRC Press: Boca Raton, FL, USA, 2012.
3. Snyder, G.J.; Toberer, E.S. Complex thermoelectric materials. *Nat. Mater.* **2008**, *7*, 105–114. [CrossRef] [PubMed]
4. Brown, S.R.; Kauzlarich, S.M.; Gascoin, F.; Snyder, G.J. $Yb_{14}MnSb_{11}$: New high efficiency thermoelectric material for power generation. *Chem. Mater.* **2006**, *18*, 1873–1877. [CrossRef]
5. Zhang, H.; Borrmann, H.; Oeschler, N.; Candolfi, C.; Schnelle, W.; Schmidt, M.; Burkhardt, U.; Baitinger, M.; Zhao, J.T.; Grin, Y. Atomic interactions in the *p*-type clathrate I $Ba_8Au_{5.3}Ge_{40.7}$. *Inorg. Chem.* **2011**, *50*, 1250–1257. [CrossRef] [PubMed]
6. Shi, X.; Yang, J.; Bai, S.; Yang, J.; Wang, H.; Chi, M.; Salvador, J.R.; Zhang, W.; Chen, L.; Wong-Ng, W. On the design of high-efficiency thermoelectric clathrates through a systematic cross-substitution of framework elements. *Adv. Funct. Mater.* **2010**, *20*, 755–763. [CrossRef]
7. Toberer, E.S.; Zevalkink, A.; Crisosto, N.; Snyder, G.J. The zintl compound $Ca_5Al_2Sb_6$ for low-cost thermoelectric power generation. *Adv. Funct. Mater.* **2010**, *20*, 4375–4380. [CrossRef]
8. Gougeon, P.; Gall, P.; Al Rahal Al Orabi, R.; Fontaine, B.; Gautier, R.; Potel, M.; Zhou, T.; Lenoir, B.; Colin, M.; Candolfi, C.; et al. Synthesis, crystal and electronic structures, and thermoelectric properties of the novel cluster compound $Ag_3In_2Mo_{15}Se_{19}$. *Chem. Mater.* **2012**, *24*, 2899–2908. [CrossRef]
9. Al Rahal Al Orabi, R.; Gougeon, P.; Gall, P.; Fontaine, B.; Gautier, R.; Colin, M.; Candolfi, C.; Dauscher, A.; Hejtmanek, J.; Malaman, B.; et al. X-ray characterization, electronic band structure, and thermoelectric properties of the cluster compound $Ag_2Tl_2Mo_9Se_{11}$. *Inorg. Chem.* **2014**, *53*, 11699–11709. [CrossRef] [PubMed]
10. Kurosaki, K.; Yamanaka, S. Low-thermal-conductivity group 13 chalcogenides as high-efficiency thermoelectric materials. *Phys. Status Solidi A* **2013**, *210*, 82–88. [CrossRef]
11. Lu, X.; Morelli, D.T.; Xia, Y.; Zhou, F.; Ozolins, V.; Chi, H.; Zhou, X.; Uher, C. High performance thermoelectricity in earth-abundant compounds based on natural mineral tetrahedrites. *Adv. Energy Mater.* **2013**, *3*, 342–348. [CrossRef]
12. Suekuni, K.; Kim, F.S.; Nishiate, H.; Ohta, M.; Tanaka, H.I.; Takabatake, T. High-performance thermoelectric minerals: Colusites $Cu_{26}V_2M_6S_{32}$ (M = Ge, Sn). *Appl. Phys. Lett.* **2014**, *105*, 132107. [CrossRef]
13. Zhang, H.; Liu, C.X.; Qi, X.L.; Dai, X.; Fang, Z.; Zhang, S.C. Topological insulators in Bi_2Se_3, Bi_2Te_3 and Sb_2Te_3 with a single dirac cone on the surface. *Nature Phys.* **2009**, *5*, 438–442. [CrossRef]
14. Xia, Y.; Qian, D.; Hsieh, D.; Wray, L.; Pal, A.; Lin, H.; Bansil, A.; Grauer, D.; Hor, Y.S.; Cava, R.J.; et al. Observation of a large-gap topological-insulator class with a single dirac cone on the surface. *Nat. Phys.* **2009**, *5*, 398–402. [CrossRef]
15. Ando, Y. Topological insulator materials. *J. Phys. Soc. Jpn.* **2013**, *82*, 102001. [CrossRef]
16. Paglione, J.; Butch, N.P. Growth and characterization of topological insulators. In *Topological Insulators, Fundamentals and Perspectives*; Ortmann, F., Roche, S., Valenzuela, S.O., Eds.; Wiley-VCH Verlag GmbH & Co., KGaA: Weinheim, Germany, 2015; pp. 245–262.

17. Abrikosov, N.K.; Bankina, V.F.; Kolomoets, L.A.; Dzhaliashvili, N.V. Deviation of the solid solution from stoichiometry in the section Bi_2Te_3–Sb_2Te_3 in the region of $Bi_{0.5}Sb_{1.5}Te_3$ composition. *Izv. Akad. Nauk. SSSR Neorg. Mater.* **1977**, *13*, 827–829. (In Russian).

18. Barash, A.S.; Zhukova, T.B.; Parparov, E.Z. Structure and thermoelectric properties of $Bi_2Te_{3-x}Se_x$ and $Bi_{0.25}Sb_{1.48}Te_3$. *Izv. Akad. Nauk. SSSR Neorg. Mater.* **1976**, *12*, 1552. (In Russian)

19. Rosi, F.D.; Abeles, B.; Jensen, R.V. Materials for thermoelectric refrigeration. *J. Phys. Chem. Solids* **1959**, *10*, 191–200. [CrossRef]

20. Susmann, H.; Loof, K. Copper doping and dislocations in the system Bi_2Te_3–Sb_2Te_3. *Phys. Stat. Sol. A* **1976**, *37*, 467–471. (In German).

21. Volotskii, M.P. Investigation of the complex structure of band edges and of the mechanism of carrier scattering in Bi-Sb-Te single crystals. *Fiz. Tekh. Polupr. SSSR* **1974**, *8*, 1044. (In Russian).

22. Yim, W.M.; Amith, A. Bi–Sb alloys for magneto-thermoelectric and thermomagnetic cooling. *Solid-State Electron.* **1972**, *15*, 1141–1144. [CrossRef]

23. Caillat, T.; Carle, M.; Perrin, D.; Scherrer, H.; Scherrer, S. Study of the Bi–Sb–Te phase diagram. *J. Phys. Chem. Solids* **1992**, *53*, 227–232. [CrossRef]

24. Caillat, T.; Carle, M.; Pierrat, P.; Scherrer, H.; Scherrer, S. Thermoelectric properties of $(Bi_xSb_{1-x})_2Te_3$ single crystal solid solutions grown by the THM method. *J. Phys. Chem. Solids* **1992**, *53*, 1121–1129. [CrossRef]

25. Qinghui, J.; Junyou, Y.; Yong, L.; Hongcai, H. Microstructure tailoring in nanostructured thermoelectric materials. *J. Adv. Dielect.* **2016**, *6*, 1630002.

26. Fan, X.A.; Yang, J.Y.; Chen, R.G.; Yun, H.S.; Zhu, W.; Bao, S.Q.; Duan, X.K. Characterization and thermoelectric properties of *p*-type 25%Bi_2Te_3–75%Sb_2Te_3 prepared via mechanical alloying and plasma activated sintering. *J. Phys. D Appl. Phys.* **2006**, *39*, 740–745. [CrossRef]

27. Vasilevskiy, D.; Dawood, M.S.; Masse, J.P.; Turenne, S.; Masut, R.A. Generation of nanosized particles during mechanical alloying and their evolution through the hot extrusion process in bismuth-telluride-based alloys. *J. Electron. Mater.* **2010**, *39*, 1890–1896. [CrossRef]

28. Navratil, J.; Starý, Z.; Plechacek, T. Thermoelectric properties of *p*-type antimony bismuth telluride alloys prepared by cold pressing. *Mater. Res. Bull.* **1996**, *31*, 1559–1566. [CrossRef]

29. Yamashita, O.; Tomiyoshi, S.; Makita, K. Bismuth telluride compounds with high thermoelectric figures of merit. *J. Appl. Phys.* **2003**, *93*, 368–374. [CrossRef]

30. Pierrat, P.; Dauscher, A.; Lenoir, B.; Martin-Lopez, R.; Scherrer, H. Preparation of the $Bi_8Sb_{32}Te_{60}$ solid solution by mechanical alloying. *J. Mater. Sci.* **1997**, *32*, 3653–3657. [CrossRef]

31. Martin-Lopez, R.; Lenoir, B.; Dauscher, A.; Scherrer, H.; Scherrer, S. Preparation of *n*-type Bi–Sb–Te thermoelectric material by mechanical alloying. *Solid State Commun.* **1998**, *108*, 285–288. [CrossRef]

32. Glazov, V.M.; Yatmanov, Y.V. Thermoelectric properties of $Bi_2Te_{2.4}Se_{0.6}$ and $Bi_{0.52}Sb_{1.48}Te_3$ semiconductor solid solutions prepared by ultrarapid liquid quenching. *Izv. Akad. Nauk. SSSR Neorg. Mater.* **1986**, *22*, 36–40. (In Russian)

33. Glazov, V.M.; Potemkin, A.Y.; Akopyan, R.A. Preparation of homogeneous inorganic solid solutions by diffusionless solidification. *Izv. Akad. Nauk. SSSR Neorg. Mater.* **1996**, *32*, 1461–1465. (In Russian).

34. Gogishvili, O.S.; Lalikin, S.P.; Krivoruchko, S.P.; Pyrychidi, K.I.; Zanava, E.S. Synthesis of alloys based on chalcogenides of Bi and Sb by ultra-fast quenching method. In Proceedings of the VII Chemistry, Physics and Technical Application of Chalcogenides, Uzhhorod, Ukraine, 24–27 October 1988; p. 368. (In Russian)

35. Gogishvili, O.S.; Kononov, G.G.; Krivoruchko, S.P.; Lavrinenko, I.P.; Ovsyanko, I.I. Structural study of $(Bi,Sb)_2Te_3$ alloys synthesized by quenching from the liquid state. *Izv. Akad. Nauk SSSR Neorg. Mater.* **1991**, *27*, 923. (In Russian)

36. Dreglea, A. *Boundary-Value Problems in Melt Spinning Modeling: Analytical and Numerical Methods*; Lambert Acad. Publ. GmbH & Co., KG: Saarbrucken, Germany, 2012.

37. Xie, W.; Tang, X.; Yan, Y.; Zhang, Q.; Tritt, T.M. High thermoelectric performance BiSbTe alloy with unique low-dimensional structure. *J. Appl. Phys.* **2009**, *105*, 113713. [CrossRef]

38. Xie, W.; Tang, X.; Yan, Y.; Zhang, Q.; Tritt, T.M. Unique nanostructures and enhanced thermoelectric performance of melt-spun BiSbTe alloys. *Appl. Phys. Lett.* **2009**, *94*, 102111. [CrossRef]

39. Xie, W.; He, J.; Kang, H.J.; Tang, X.; Zhu, S.; Laver, M.; Wang, S.; Copley, J.R.D.; Brown, C.M.; Zhang, Q.; et al. Identifying the specific nanostructures responsible for the high thermoelectric performance of $(Bi,Sb)_2Te_3$ nanocomposites. *Nano Lett.* **2010**, *10*, 3283–3289. [CrossRef] [PubMed]

40. Xie, W.J.; He, J.; Zhu, S.; Su, X.L.; Wang, S.Y.; Holgate, T.; Graff, J.W.; Ponnambalam, V.; Poon, S.J.; Tang, X.F.; et al. Simultaneously optimizing the independent thermoelectric properties in (Ti,Zr,Hf)(Co,Ni)Sb alloy by in situ forming InSb nanoinclusions. *Acta Mater.* **2010**, *58*, 4705–4713. [CrossRef]

41. Xie, W.; He, J.; Zhu, S.; Holgate, T.; Wang, S.; Tang, X.; Zhang, Q.; Tritt, T.M. Investigation of the sintering pressure and thermal conductivity anisotropy of melt-spun spark-plasma-sintered (Bi,Sb)$_2$Te$_3$ thermoelectric materials. *J. Mater. Res.* **2011**, *26*, 1791–1799. [CrossRef]

42. Xie, W.; Wang, S.; Zhu, S.; He, J.; Tang, X.; Zhang, Q.; Tritt, T.M. High performance Bi$_2$Te$_3$ nanocomposites prepared by single-element-melt-spinning spark-plasma sintering. *J. Mater. Sci.* **2013**, *48*, 2745–2760. [CrossRef]

43. Ivanova, L.D.; Petrova, L.I.; Granatkina, Y.V.; Leontyev, V.G.; Ivanov, A.S.; Varlamov, S.A.; Prilepo, Y.P.; Sychev, A.M.; Chuik, A.G.; Bashkov, I.V. Thermoelectric and mechanical properties of the Bi$_{0.5}$Sb$_{1.5}$Te$_3$ solid solution prepared by melt spinning. *Inorg. Mater.* **2013**, *49*, 120–126. [CrossRef]

44. Koukharenko, E.; Frety, N.; Shepelevich, V.G.; Tedenac, J.C. Thermoelectric properties of Bi$_2$Te$_3$ material obtained by the ultrarapid quenching process route. *J. Alloys Compd.* **2000**, *299*, 254–257. [CrossRef]

45. Koukharenko, E.; Vassilev, G.P.; Nancheva, N.; Docheva, P.; Tedenac, J.C.; Frety, N.; Shepelevich, V.G. defects in Sb$_{2−x}$Bi$_x$Te$_3$ foils. *J. Alloys Compd.* **1999**, *287*, 239–242. [CrossRef]

46. Koukharenko, E.; Frety, N.; Nabias, G.; Shepelevich, V.G.; Tedenac, J.C. Microstructural study of Bi$_2$Te$_3$ material obtained by ultrarapid quenching process route. *J. Cryst. Growth* **2000**, *209*, 773–778. [CrossRef]

47. Koukharenko, E.; Shepelevich, V.G. Structural and thermoelctric properties of rapidly quenched Bi$_{2−x}$Sb$_x$Te foils. *Inorg. Mater.* **1999**, *35*, 115–117.

48. Koukharenko, E.; Frety, N.; Shepelevich, V.G.; Tedenac, J.C. Microstructure and thermoelectric properties of thin foils of bismuth telluride alloys. *Mater. Res. Soc. Symp. Proc.* **1999**, *545*, 507–512. [CrossRef]

49. Koukharenko, E.; Frety, N.; Shepelevich, V.G.; Tedenac, J.C. Electrical and microstructural properties of Bi$_{2−x}$Sb$_x$Te and Bi$_{2−x}$Sb$_x$Te$_2$ foils obtained by the ultrarapid quenching process. *J. Mater. Sci. Mater. Electron.* **2003**, *14*, 383–388. [CrossRef]

50. Ohorodniichuk, V.; Candolfi, C.; Masschelein, P.; Baranek, P.; Dalicieux, P.; Dauscher, A.; Lenoir, B. Influence of preparation processing on the transport properties of melt-spun Sb$_{2−x}$Bi$_x$Te$_{3+y}$. *J. Electron. Mater.* **2016**, *45*, 1561–1569. [CrossRef]

51. Huang, S.C.; Laforce, R.; Ritter, A.; Goehner, R. Rapid solidification characteristics in melt-spinning a Ni-base superalloy. *Met. Trans. A* **1985**, *16*, 1773–1779. [CrossRef]

52. Fedotov, A.S.; Svito, I.A.; Gusakova, S.V.; Shepelevich, V.G.; Saad, A.; Mazanik, A.V.; Fedotov, A.K. Electronic properties of Bi-Sn diluted alloys. *Mater. Today* **2014**, *2*, 629–636. [CrossRef]

53. Boettinger, W.J.; Perepezko, J.H. *Rapidly Solidified Alloys: Processes, Structures, Properties, Applications*; Liebermann, H.H., Ed.; Marcel Dekker Inc.: New York, NY, USA, 1993.

54. Birkholz, M. *Thin Film Analysis by X-ray Scattering*; John Wiley & Sons: Somerset, NJ, USA, 2006.

55. Creagh, D.C.; Hubbell, J.H. X-ray absorption (or attenuation) coefficients. In *International Tables for Crystallography C*; Wilson, A.J.C., Prince, E., Kluwer, D., Eds.; Kluwer Academic Publishers: Dordrecht, The Netherlands, 1999; p. 220.

56. Almeida, M.; Alcácer, L.; Oostra, S. Anisotropy of Thermopower in N-methyl-N-ethylmorpholinium bistetracyanoquinodimethane, MEM(TCNQ)$_2$ in the Region of the High-Temperature Phase Transitions. *Phys. Rev. B* **1984**, *30*, 2839–2844. [CrossRef]

57. Huebener, H.P. Thermoelectric Power of Lattice Vacancies of Gold. *Phys. Rev.* **1964**, *135*, A1281–A1291. [CrossRef]

58. Ohorodniichuk, V.; Dauscher, A.; Masschelein, P.; Candolfi, C.; Baranek, P.; Dalicieux, P.; Lenoir, B. Influence of the nozzle diameter as a control parameter of the properties of melt-spun Sb$_{2−x}$Bi$_x$Te$_3$. *J. Electron. Mater.* **2016**, *45*, 1419–1424. [CrossRef]

59. Ivanov, L.D.; Kop'ev, I.M.; Lobzov, M.A.; Abrikosov, N.K. Mechanical properties of sngle-crystals of solid solutions of the system Sb$_{1.5}$Bi$_{0.5}$Te$_3$–Bi$_2$Se$_3$. *Inorg. Mater.* **1987**, *23*, 1288–1291.

60. Jacquot, A.; Vollmer, F.; Bayer, B.; Jaegle, M.; Ebling, D.G.; Böttner, H. Thermal conductivity measurements on challenging samples by the 3-omega method. *J. Electron. Mater.* **2010**, *39*, 1621–1626. [CrossRef]

61. Goltsman, B.M.; Kudinov, V.A.; Smirnov, I.A. *Semiconductor Thermoelectric Materials Based on Bi$_2$Te$_3$ (Nauka, Moscow, 1972)*; Army Foreign Science and Technology Center: Charlottesville, VA, USA, 1973.

62. Lan, Y.; Poudel, B.; Ma, Y.; Wang, D.; Dresselhaus, M.S.; Chen, G.; Ren, Z. Structure study of bulk nanograined thermoelectric bismuth antimony telluride. *Nano Lett.* **2009**, *9*, 1419–1422. [CrossRef] [PubMed]

63. Li, G.; Gadelrab, K.R.; Souier, T.; Potapov, P.L.; Chen, G.; Chiesa, M. Mechanical properties of $Bi_xSb_{2-x}Te_3$ nanostructured thermoelectric material. *Nanotechnology* **2012**, *23*, 065703. [CrossRef] [PubMed]

64. Peranio, N.; Eibl, O. Structural modulations in Bi_2Te_3. *J. Appl. Phys.* **2008**, *103*, 024314. [CrossRef]

65. Homer, M.D.; Medlin, L. Preparation Methods for TEM Specimens of Bismuth Telluride and Related Thermoelectric Alloys. *Microsc. Microanal.* **2012**, *18*, 1482–1500. [CrossRef]

66. Thonhauser, T.; Jeon, G.S.; Mahan, G.D.; Sofo, J.O. Stress-induced defects in Sb_2Te_3. *Phys. Rev. B* **2003**, *68*, 205207. [CrossRef]

67. Pecheur, P.; Toussaint, G. Tight-binding studies of crystal stability and defects in Bi_2Te_3. *J. Phys. Chem. Solids* **1994**, *55*, 327–338. [CrossRef]

68. Manyakin, S.M.; Volkov, M.P. Microstructure of $(Bi_{0.25}Sb_{0.75})_2Te_3$ Profiled Crystals Grown by Directed Crystallization Method. In Proceedings of the Twenty-First International Conference on Thermoelectrics, Long Beach, CA, USA, 25–29 August 2002; pp. 21–23.

© 2017 by the authors. Licensee MDPI, Basel, Switzerland. This article is an open access article distributed under the terms and conditions of the Creative Commons Attribution (CC BY) license (http://creativecommons.org/licenses/by/4.0/).

crystals

MDPI

Article

Synthesis and Thermoelectric Properties of Copper Sulfides via Solution Phase Methods and Spark Plasma Sintering

Yun-Qiao Tang, Zhen-Hua Ge * and Jing Feng

Faculty of Materials Science and Engineering, Kunming University of Science and Technology, Kunming 650093, China; yqt-joe@foxmail.com (Y.-Q.T.); jingfeng@kmust.edu.cn (J.F.)
* Correspondence: zge@kmust.edu.cn

Academic Editor: George S. Nolas
Received: 27 April 2017; Accepted: 13 May 2017; Published: 16 May 2017

Abstract: Large-scale Cu_2S tetradecahedrons microcrystals and sheet-like Cu_2S nanocrystals were synthesized by employing a hydrothermal synthesis (HS) method and wet chemistry method (WCM), respectively. The morphology of α-Cu_2S powders prepared by the HS method is a tetradecahedron with the size of 1–7 μm. The morphology of β-Cu_2S is a hexagonal sheet-like structure with a thickness of 5–20 nm. The results indicate that the morphologies and phase structures of Cu_2S are highly dependent on the reaction temperature and time, even though the precursors are the exact same. The polycrystalline copper sulfides bulk materials were obtained by densifying the as-prepared powders using the spark plasma sintering (SPS) technique. The electrical and thermal transport properties of all bulk samples were measured from 323 K to 773 K. The pure Cu_2S bulk samples sintered by using the powders prepared via HS reached the highest thermoelectric figure of merit (ZT) value of 0.38 at 573 K. The main phase of the bulk sample sintered by using the powder prepared via WCM changed from β-Cu_2S to $Cu_{1.8}S$ after sintering due to the instability of β-Cu_2S during the sintering process. The $Cu_{1.8}S$ bulk sample with a $Cu_{1.96}S$ impurity achieved the highest ZT value of 0.62 at 773 K.

Keywords: Cu_2S; nanocrystal; synthetic methods; morphological control; thermoelectric properties

---

1. Introduction

Thermoelectric (TE) material is a kind of energy conversion material which takes advantage of solid material internal carriers and phonon interactions to convert thermal and electrical energy directly into each other. The energy crisis and environmental problems have promoted the swift development of TE materials in the past few decades [1–4]. Compared to the mainstream tellurium-based TE materials [5–8], nanostructured metal chalcogenides with low cost, low toxicity, and abundant elements exhibit interesting physical properties [9,10]. Therefore, nanostructured metal chalcogenides TE materials such as Cu-Se [11–13] and Cu-S materials [14–17] have received more attention.

Copper sulfides ($Cu_{2-x}S$ ($0 \leq x \leq 1$)), with different copper stoichiometric ratios, which are a series of compounds ranging from copper-rich Cu_2S to copper deficient CuS, are considered to be superionic conductors [18]. As an important semiconductor, Cu_2S is of high interest due to its unique electronic, thermodynamic, optical, and other physical and chemical properties. It has great potential in a wide range of applications such as thermoelectric materials [19], solar cells [20,21], conductive fibers [22], optical filters [23], and high-capacity cathode materials in lithium secondary batteries [24]. Moreover, Cu_2S nanoparticles with various morphologies have been synthesized by various approaches such as chemical precipitation [25], solventless thermolysis [26], water-oil interface confined method [27], and thermal decomposition [28].

Here, we employed facile solution methods, including hydrothermal synthesis (HS) and wet chemistry method (WCM), to synthesize Cu_2S powders with controllable microstructures under relatively facile conditions. Then, polycrystalline copper sulfides were fabricated by densifying the compound powders using the spark plasma sintering (SPS) technique. The thermoelectric properties of all the bulk samples were measured.

2. Experimental Section

Commercial high-purity powders of CuO (99.9%) and S (99.99%) were used as raw materials. Meanwhile, ethylene diamine (EDA) and hydrazine hydrate ($N_2H_4 \cdot H_2O$) were used as a chelating agent (EDA) and a reducing agent ($N_2H_4 \cdot H_2O$), respectively. In a typical wet chemistry method, CuO (10 mmol) and S (20 mmol) were first added to EDA (40 mL) by stirring at room temperature for 10 min. Then, $N_2H_4 \cdot H_2O$ (35 mL) was dripped slowly into the beaker under further stirring for 12 h at room temperature. The chelating agent EDA reacted with Cu ions to form the complex compounds for avoiding the precipitation of metal Cu. The reducing agent $N_2H_4 \cdot H_2O$ reduced the Cu^{2+} to Cu^+ and S to S^{2-}, respectively.

In a typical hydrothermal synthesis [29] method, CuO (10 mmol) and S (20 mmol) were first added to EDA (40 mL), and the mixture was stirred and heated to 373 K for 10 min. After that, $N_2H_4 \cdot H_2O$ (35 mL) was dripped slowly into the solution under further stirring for 10 min at 373 K. The mixed solution was then transferred into a Teflon-lined stainless steel autoclave (100 mL capacity), which was sealed and maintained at 453 K for 6 h. The final solid products were filtered and washed with DI water and ethanol three times before drying under vacuum at 333 K for 12 h.

The resultant powders were loaded into a graphite die with an inner diameter of 15 mm and then sintered at 773 K for 5 min (heating rate of 100 K/min) under an axial compressive stress of 40 MPa in a vacuum by using a spark plasma sintering (SPS) system (SPS1050; Sumitomo, Tokyo, Japan). The SPS-prepared specimens were disk-shaped with dimensions of $\Phi15$ mm × 4 mm. The phase structure was analyzed by X-ray diffraction with a Cu K_α radiation ($\lambda = 1.5406$ Å) filtered through Ni foil (RAD-B system; Rigaku, Tokyo, Japan). The morphologies of the powders and the fracture of the bulk samples were observed by field-emission scanning electron microscopy (FESEM, SUPRA 55, Carl Zeiss, Oberkochen, Germany). The microstructure of the powder was also checked using transmission electron microscopy (TEM, Phililp Tecnai F20, Amsterdam, Dutch). In a typical TEM sample preparation procedure, powders were first added to ethyl alcohol, and stirred for 10 min by ultrasound. Then, the supernatant was dropped on the copper grid.

The electrical transport properties were evaluated along a sample section perpendicular to the SPS pressing direction. The Seebeck coefficient and electrical resistivity were measured from 323 to 773 K in a helium atmosphere using a Seebeck coefficient/electrical resistance measuring system (ZEM-3, Ulvac-Riko, Kanagawa, Japan). The density (d) of the sample was measured by the Archimedes method. In addition, the thermal conductivity of the samples was calculated by the relationship $\kappa = DC_pd$ from the thermal diffusivity D measured by the laser flash method (LFA457; NETZSCH, Selb, Bavaria, Germany).

3. Results and Discussion

3.1. Powder Synthesis and Characterization

3.1.1. XRD Analysis

Figure 1 shows the XRD patterns of Cu_2S powders which prepared by WCM and HS methods. All of the diffraction peaks of the HS sample are well-matched with the standard card of α-Cu_2S (JSPDS no. 83-1462), showing that the pure monoclinic α-Cu_2S powders were obtained by HS methods. All of the diffraction peaks of the WCM sample are well-matched with the standard card β-Cu_2S (JCPDS no. 26-1116), showing that the pure hexagonal β-Cu_2S powders were obtained by WCM.

Actually, the hexagonal β-Cu$_2$S is high-chalcocite (378–698 K); it would be not stable under thermal shock. The diffraction peaks of powder samples are wider compared to the standard card, indicating the small grain size of the obtained powder. The XRD refinement was performed for the two samples as shown in Figures S1 and S2. The location, proportion and lattice constant of experiment and refinement Cu ions for hexagonal β-Cu$_2$S powder are shown in Table S1. The proportion of Cu1 changed from 0.75 to 0.4896, indicating that there are more Cu vacancies, smaller lattice parameters, and the possibility of defects in hexagonal Cu$_2$S.

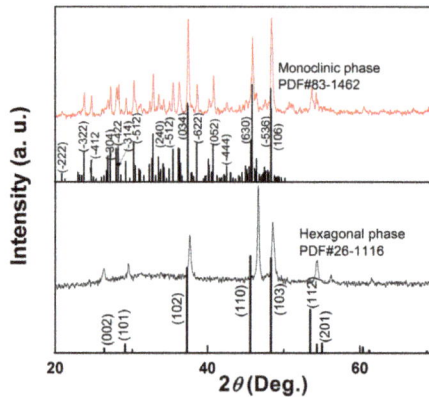

Figure 1. XRD patterns with a selected 2θ range of 20°–70° for Cu$_2$S powders.

3.1.2. FESEM and TEM Analysis

Figure 2 shows the FESEM images of Cu$_2$S prepared by different synthesis methods. The pure α-Cu$_2$S powder which was prepared by the HS method is shown in Figure 2a. The morphology of α-Cu$_2$S powder prepared by the HS method is a tetradecahedron with the size of 1–7 μm. Figure 2b is a magnified image of a typical single-crystalline α-Cu$_2$S, shown in Figure 2a. The pure β-Cu$_2$S powder which was prepared by WCM is shown in Figure 2c. The morphology of hexagonal β-Cu$_2$S is hexagonal nanosheets. Each nanosheet has an edge length of 10–200 nm and a thickness of 5–20 nm. The morphologies of the powder are highly related to the crystal structure [30]. Similar work reported in the literature for molybdate materials [31] suggests that the synthesis temperature has a strong influence on the morphology of Cu$_2$S samples. Under the synthesis conditions of high temperature and high pressure, the monoclinic α-Cu$_2$S showed a tetradecahedron morphology. Also, under relatively mild synthesis conditions (WCM), the hexagonal β-Cu$_2$S nanosheets were synthesized at room temperature.

The high magnification FEM image of β-Cu$_2$S is shown in Figure 2d. The β-Cu$_2$S has a hexagonal sheet-like structure with a thickness of 20 nm. The TEM image (Figure 3a) and selected area of electron diffraction (SAED) patterns (Figure 3b) of a single crystal α-Cu$_2$S revealed a tetradecahedron Cu$_2$S of 1 μm in width. The SAED pattern for the tetradecahedron particles (Figure 2b) indicated a highly crystallized monoclinic structure of the α-Cu$_2$S. Additionally, the TEM image (Figure 3c) and SAED patterns (Figure 3d) of a single-crystalline β-Cu$_2$S revealed a sheet-like Cu$_2$S of 10–200 nm in width and of 5–20 nm in thickness. The results are in agreement with SEM observations.

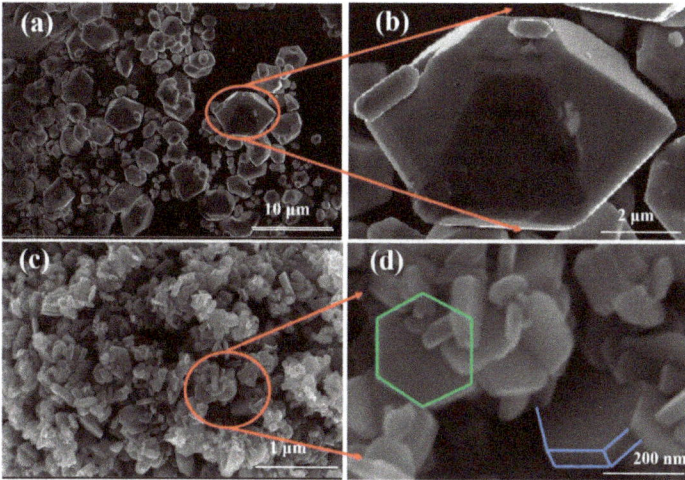

Figure 2. Field emission scanning electron microscopy patterns of the powders under different magnifications. (**a,b**) Cu_2S powders prepared by HS; (**c,d**) Cu_2S powders prepared by WCM.

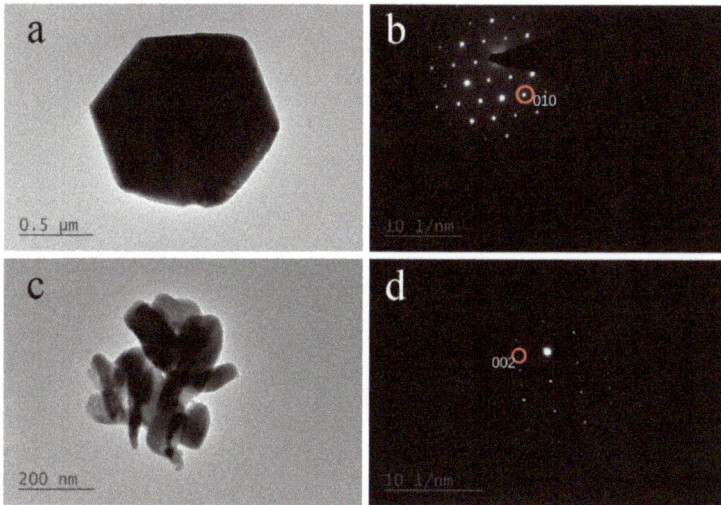

Figure 3. TEM image and SAED patterns for HS-Cu_2S powder (**a,c**) and WCM-Cu_2S powder (**b,d**).

3.1.3. Synthesis Mechanism

The synthesis process and mechanism of Cu_2S, similar to the flower-like α-Fe_2O_3 reported by Penki et al. [32], were investigated in detail, as shown in Figure 4. The raw CuO (black) and S (yellow) were mixed in a $N_2H_4 \cdot H_2O$ and EDA solution to initially produce the precursor Cu_2O and S^{2-}, and then further react to become the product Cu_2S under the different reaction conditions. When the HS method was employed, the reaction temperature was 453 K and the product was monoclinic α-Cu_2S, as indicated in XRD shown in Figure 1. After 6 h of hydrothermal reaction, the single crystal monoclinic α-Cu_2S formed a tetradecahedron with a dimension of several micrometers as shown in Figure 2a. But when WCM was employed with a reaction temperature of room temperature, the product was hexagonal β-Cu_2S, as indicated in XRD shown in Figure 1. After a 12-h reaction, the single crystal hexagonal β-Cu$_2$S grew to form nanosheets with an edge length of 10–200 nm and thickness of 5–20 nm as shown in Figure 2c. The reaction process can be described by Equations (1)–(3).

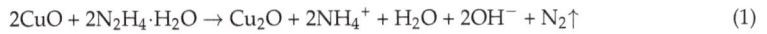

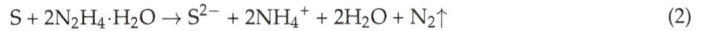

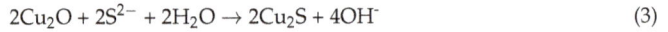

$$2CuO + 2N_2H_4 \cdot H_2O \rightarrow Cu_2O + 2NH_4^+ + H_2O + 2OH^- + N_2\uparrow \tag{1}$$

$$S + 2N_2H_4 \cdot H_2O \rightarrow S^{2-} + 2NH_4^+ + 2H_2O + N_2\uparrow \tag{2}$$

$$2Cu_2O + 2S^{2-} + 2H_2O \rightarrow 2Cu_2S + 4OH^- \tag{3}$$

Finally, it is suggested that the morphologies and phase structures of Cu$_2$S are highly dependent on the reaction temperature and time, even if precursors are exactly the same.

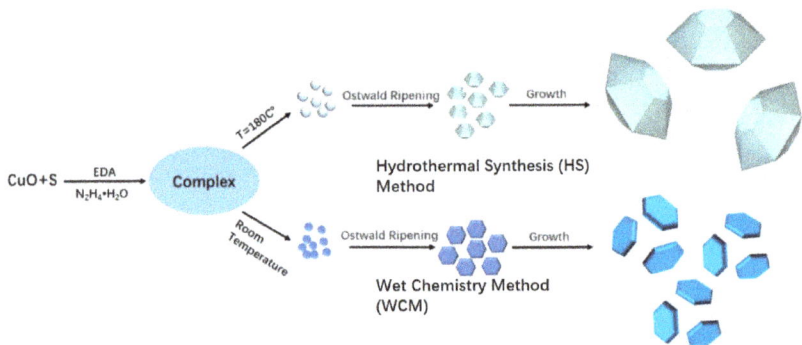

Figure 4. Schematic illustration of the growth mechanism of the Cu$_2$S powders synthesized by the WCM and HS methods.

3.2. Bulk Characterization

3.2.1. XRD and FESEM Analysis

The bulk sample abbreviated as HS-bulk was prepared by applying SPS at 773 K for 5 min using monoclinic α-Cu$_2$S powders. The bulk sample abbreviated as WCM-bulk was prepared by applying SPS at 773 K for 5 min using hexagonal β-Cu$_2$S powders. The XRD patterns of those bulk samples are shown in Figure 5a. The green arrows index the impurity peaks of Cu$_{1.96}$S. The HS-bulk is still in the α-Cu$_2$S phase. The SEM image of HS-bulk shown in Figure 5b shows a high relative density of 96% and an average grain size of 5 μm, which is very similar to the powders formed by the rapid sintering process. SPS is a rapid sintering technology, after which the nanoscale particles can be maintained in bulk [14]. As shown in the XRD pattern (Figure 5a), the WCM-bulk exhibits a main phase of Cu$_{1.8}$S with an impurity of Cu$_{1.96}$S. The hexagonal β-Cu$_2$S is in the high-chalcocite phase, which is instable during the sintering process. Because of thermal shock in the SPS process, the phase transition occurred from β-Cu$_2$S to Cu$_{1.8}$S, which is the most stable phase in the Cu-S system. Cu$_2$S is a superionic conductor, and the superionic phase transition temperature is over 689 K [33]. In the SPS sintering program, the sintering temperature is 773 K, and the Cu ions show liquid behavior under this temperature. The extra Cu may precipitate on the anode and be removed during the polishing and cutting process. As shown in Figure 5c, the WCM-bulk sample with a relative density of 90.2% has an average size of 200 nm.

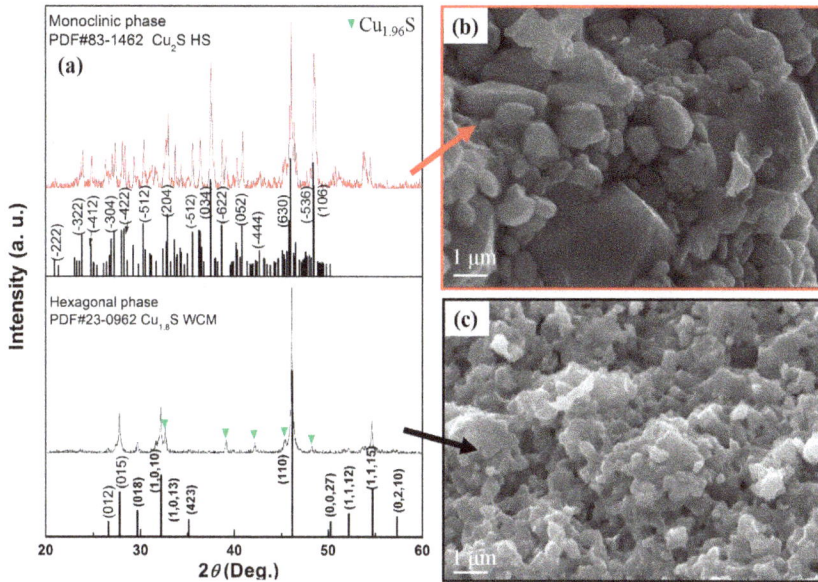

Figure 5. XRD patterns with a selected 2θ range of 20°–70° for bulk samples prepared by SPS at 773 K for 5 min, and field emission scanning electron microscopy of the fractured surfaces for the bulk samples. (**a**) XRD patterns of the two bulk samples; (**b**) FESEM image of HS-bulk; (**c**) FESEM image of WCM-bulk.

3.2.2. Thermoelectric Transport Properties

The TE properties of the two bulk samples were measured. The TE properties of $Cu_{1.8}S$ bulk sample in the literature [34], which were prepared by the same SPS process but using the mechanical alloying (MA) treated powders, are shown for comparison. Figure 6a illustrates the temperature dependence of electrical conductivity (σ) for the bulk samples. As shown in Figure 6a, the HS-bulk has the highest σ value up to 218 Scm^{-1} at 373 K, and WCM-bulk has the highest σ value up to 2490 Scm^{-1} at 373 K. Both the HS-bulk and the WCM-bulk have lower σ than the MA-bulk. The WCM-bulk has a similar main phase to the MA-bulk but a low relative density and impurity of $Cu_{1.96}S$, which decrease the σ value. The HS-bulk sample has the main phase of Cu_2S, while the electrical conductivity of $Cu_{2-x}S$ depends on the Cu content due to its superionic behavior. Therefore, the HS-bulk sample has the lowest electrical conductivity. The WCM-bulk sample has one turning point in the σ curve due to the one phase transition of $Cu_{1.8}S$ during the measured temperature range of 323 K to 773 K [34]. Two turning points in the σ curve were observed for the HS-bulk sample due to the two phase transitions of Cu_2S during the measured temperature range. As reported by Li et al., bulk Cu_2S exhibits three phases (α-phase, β-phase, γ-phase) in the temperature ranges of >698, 378–698, and <378 K [35,36], respectively. The high temperature region to the right of the blue dashed line in Figures 6 and 7 is the γ-phase.

The positive Seebeck coefficient (α) in Figure 6b indicates all bulks are p-type semiconductors. According to the equations $\sigma = e\mu n$, and $\alpha \approx \gamma - \ln n$, [37] where σ, μ, n, α and γ are electrical conductivity, carrier mobility, carrier concentration, Seebeck coefficient and scattering factor, respectively, α is usually inversely proportional to σ. The HS-bulk achieved the largest α value of 532 μVK^{-1} at 673 K. The WCM-bulk achieved the largest α value of 101 μVK^{-1} at 773 K. The *PF* was calculated by $PF = \alpha^2\sigma$ and is shown in Figure 6c. The *PF* of the HS-bulk reaches 196 μWm^{-1}K^{-2} at 573 K, and that of the WCM-bulk achieved 985 μWm^{-1}K^{-2} at 773 K.

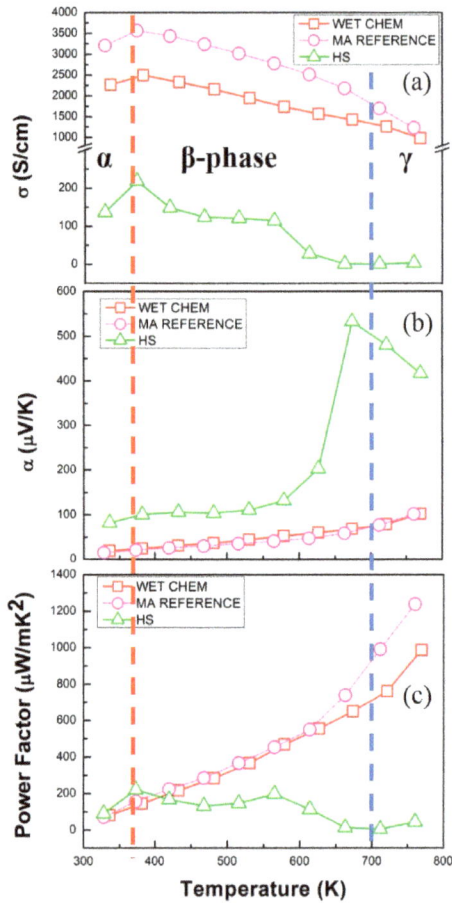

Figure 6. Temperature dependence of electrical conductivity (**a**); Seebeck coefficient (**b**); and power factor (**c**) for the two bulk samples and the reference [34].

Figure 7 shows the temperature dependence of thermal conductivity (κ) (a) and the thermoelectric figure of merit, ZT (b). The κ value of the HS-bulk and the WCM-bulk are lower than that of the MA-bulk [34], due to the fine grain size and the lower relative density. The HS-bulk sample obtained the lowest κ value of 0.20 WmK^{-1} at 673 K. The WCM-bulk sample obtained the lowest κ value of 1.23 WmK^{-1} at 773 K. The κ curve of the HS sample also has two turning points according to the two phases transitions of Cu$_2$S, which are also similar to the previous report by He et al. [33]. Based on the above measurement results, the ZT was calculated by $ZT = \sigma \alpha^2 \, T / \kappa$ as shown in Figure 7b. The highest ZT value of 0.38 was obtained at 573 K for the HS-bulk sample, and the WCM-bulk sample obtained the highest ZT value of 0.62 at 773 K. This shows that Cu$_2$S is a promising thermoelectric material, and the method combining the solution phase method and SPS may be an efficient route for synthesizing high performance bulk TE materials [37].

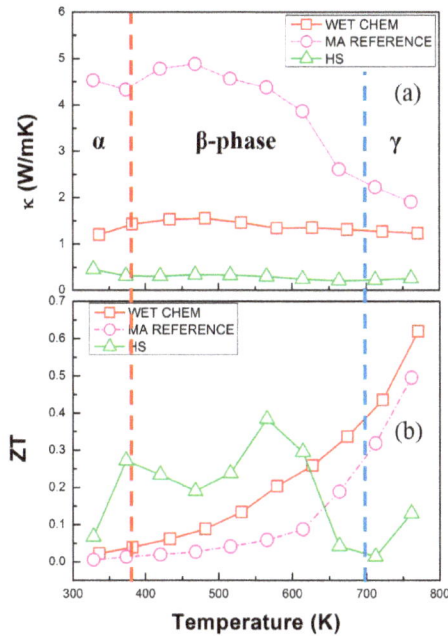

Figure 7. Temperature dependence of thermal conductivity (**a**) for the two bulk samples and *ZT* value and (**b**) for the two bulk samples and the reference [34].

4. Conclusions

The stable solution phase processes for preparing the single phase Cu_2S powders were obtained. Large-scale Cu_2S tetradecahedron microcrystals with monoclinic symmetry and sheet-like Cu_2S nanocrystals with hexagonal β-Cu_2S symmetry were synthesized by employing the hydrothermal synthesis (HS) method and the wet chemistry method (WCM), respectively. The Cu_2S nanopowders were densified to bulk by SPS. Due to the fine grains, low thermal conductivity was achieved, resulting in enhanced TE properties. The highest ZT value of 0.38 was obtained at 573 K for the HS-bulk sample, which is better than the values achieved by the other two samples in this temperature. The WCM-bulk sample obtained the highest ZT value of 0.62 at 773 K. Compared with the MA-bulk [34], the WCM-bulk has a 26.53% increase of ZT value at 773 K. Our work indicated that the morphologies and phase structures of Cu_2S are highly dependent on the reaction temperature and time even when the raw materials were exactly the same.

Supplementary Materials: The following are available online at http://www.mdpi.com/2073-4352/7/5/141/s1, Figure S1: XRD refinement patterns of the hexagonal Cu2S powder sample, Figure S2: XRD refinement patterns of the monoclinic Cu2S powder sample, Table S1: The location, proportion and lattice constant of Cu ions, Table S2: The lattice constant of monoclinic Cu2S powder sample.

Acknowledgments: This work was supported by the National Natural Science Foundation of China (Grant No. 51501086).

Author Contributions: All authors participated in the research, analysis and writing of the manuscript. Zhen-Hua Ge designed the experiments and Yun-Qiao Tang fabricated the samples, preformed the thermoelectric properties characterization and wrote the paper.

Conflicts of Interest: The authors declare no conflicts of interest.

References

1. Sales, B.C. Smaller is cooler. *Science* **2001**, *295*, 1248–1249. [CrossRef] [PubMed]
2. Sootsman, J.R.; Chung, D.Y.; Kanatzidis, M.G. New and old concepts in thermoelectric materials. *Angew. Chem. Int. Ed.* **2009**, *48*, 8616–8639. [CrossRef] [PubMed]
3. Snyder, G.J.; Eric, S.T. Complex thermoelectric materials. *Nat. Mater.* **2008**, *7*, 105–114. [CrossRef] [PubMed]
4. Di Salvo, F.J. Thermoelectric Cooling and Power Generation. *Science* **1999**, *285*, 703–706. [CrossRef]
5. Zhao, X.B.; Ji, X.H.; Zhang, Y.H.; Zhu, T.J.; Tu, J.P.; Zhang, X.B. Bismuth telluride nanotubes and the effects on the thermoelectric properties of nanotube-containing nanocomposites. *Appl. Phys. Lett.* **2005**, *86*, 062111. [CrossRef]
6. Zhang, Y.C.; Wang, H.; Kraemer, S.; Shi, Y.F.; Zhang, F.; Snedaker, M.; Ding, K.; Moskovits, M.; Snyder, G.J.; Stucky, G.D. Surfactant-Free Synthesis of Bi_2Te_3-Te Micro-Nano Heterostructure with Enhanced Thermoelectric Figure of Merit. *ACS Nano* **2011**, *5*, 3158–3165. [CrossRef] [PubMed]
7. Heremans, J.P.; Jovovic, V.; Toberer, E.S.; Saramat, A.; Kurosaki, K.; Charoenphakdee, A.; Yamanaka, S.; Snyder, G.J. Enhancement of thermoelectric efficiency in PbTe by distortion of the electronic density of states. *Science* **2008**, *321*, 554–557. [CrossRef] [PubMed]
8. Li, Z.Y.; Li, J.F.; Zhao, W.Y.; Tan, Q.; Wei, T.R.; Wu, C.F.; Xing, Z.B. PbTe-based thermoelectric nanocomposites with reduced thermal conductivity by SiC nanodispersion. *Appl. Phys. Lett.* **2014**, *104*, 113905. [CrossRef]
9. Gao, M.R.; Xu, Y.F.; Jiang, J.; Yu, S.H. Nanostructured metal chalcogenides: Synthesis, modification, and applications in energy conversion and storage devices. *Chem. Soc. Rev.* **2013**, *42*, 2986–3017. [CrossRef] [PubMed]
10. Ge, Z.H.; Nolas, G.S. Controllable Synthesis of Bismuth Chalcogenide Core–shell Nanorods. *Cryst. Growth Des.* **2014**, *14*, 533–536. [CrossRef]
11. Liu, H.L.; Shi, X.; Xu, F.F.; Zhang, L.L.; Zhang, W.Q.; Chen, L.D.; Uher, C.; Day, T.; Snyder, G.J. Copper ion liquid-like thermoelectrics. *Nat. Mater.* **2012**, *11*, 422–425. [CrossRef] [PubMed]
12. Li, D.; Qin, X.Y.; Liu, Y.F.; Song, C.J.; Wang, L.; Zhang, J.; Xin, H.X.; Guo, G.L.; Zou, T.H.; Sun, G.L.; et al. Chemical synthesis of nanostructured Cu_2Se with high thermoelectric performance. *RSC Adv.* **2014**, *4*, 8638–8644. [CrossRef]
13. Shi, X.; Xi, L.; Fan, J.; Zhang, W.; Chen, L. Cu-Se bond network and thermoelectric compounds with complex diamondlike structure. *Chem. Mater.* **2010**, *22*, 6029–6031. [CrossRef]
14. Ge, Z.H.; Zhang, B.P.; Chen, Y.X.; Yu, Z.X.; Liu, Y.; Li, J.F. Synthesis and transport property of $Cu_{1.8}S$ as a promising thermoelectric compound. *Chem. Commun.* **2011**, *47*, 12697–12699. [CrossRef] [PubMed]
15. Dennler, G.; Chmielowski, R.; Jacob, S.; Capet, F.; Roussel, P.; Zastrow, S.; Nielsch, K.; Opahle, I.; Madsen, G.K.H. Are binary copper sulfides/selenides really new and promising thermoelectric materials? *Adv. Energy Mater.* **2014**, *4*, 1301581. [CrossRef]
16. Abdullaev, G.B.; Aliyarova, Z.A.; Zamanova, E.H.; Asadov, G.A. Investigation of the electric properties of Cu_2S single crystals. *Phys. Status Solidi* **1968**, *26*, 65–68. [CrossRef]
17. Lu, X.; Morelli, D.T.; Xia, Y.; Zhou, F.; Ozolins, V.; Chi, H.; Zhou, X.Y.; Uher, C. High Performance Thermoelectricity in Earth-Abundant Compounds Based on Natural Mineral Tetrahedrites. *Adv. Energy Mater.* **2013**, *3*, 342–348. [CrossRef]
18. Miller, T.A.; Wittenberg, J.S.; Wen, H.; Connor, S.; Cui, Y.; Lindenberg, A.M. The mechanism of ultrafast structural switching in superionic copper (I) sulphide nanocrystals. *Nat. Commun.* **2013**, *4*, 1369. [CrossRef] [PubMed]
19. El Akkad, F.; Mansour, B.; Hendeya, T. Electrical and thermoelectric properties of Cu_2Se and Cu_2S. *Mater. Res. Bull.* **1981**, *16*, 535–539. [CrossRef]
20. Rothwarf, A. The CdS/Cu_2S solar cell: Basic operation and anomalous effects. *Sol. Cells* **1980**, *2*, 115–140. [CrossRef]
21. Hall, R.B.; Meakin, J.D. The design and fabrication of high efficiency thin film CdS/Cu2S solar cells. *Thin Solid Films* **1979**, *63*, 203–211. [CrossRef]
22. Kim, S.; Kim, S.; Kim, Y.H.; Ku, B.C.; Jeong, Y. Enhancement of electrical conductivity of carbon nanotube fibers by copper sulfide plating. *Fibers Polym.* **2015**, *16*, 769–773. [CrossRef]
23. Fernandez, A.M.; Nair, M.T.S.; Nair, P.K. Chemically deposited ZnS-NiS-CuS optical filters with wide range solar control characteristics. *Mater. Manuf. Process* **1993**, *8*, 535–548. [CrossRef]

24. Lai, C.H.; Huang, K.W.; Cheng, J.H.; Lee, C.Y.; Hwang, B.J.; Chen, L.J. Direct growth of high-rate capability and high capacity copper sulfide nanowire array cathodes for lithium-ion batteries. *J. Mater. Chem.* **2010**, *20*, 6638–6645. [CrossRef]

25. Wu, S.X.; Jiang, J.; Liang, Y.G.; Yang, P.; Niu, Y.; Chen, Y.D.; Xia, J.F.; Wang, C. Chemical Precipitation Synthesis and Thermoelectric Properties of Copper Sulfide. *J. Electron. Mater.* **2017**, *46*, 2432–2437. [CrossRef]

26. Sigman, M.B.; Ghezelbash, A.; Hanrath, T.; Saunders, A.E.; Lee, F.; Korgel, B.A. Solventless synthesis of monodisperse Cu2S nanorods, nanodisks, and nanoplatelets. *J. Am. Chem. Soc.* **2003**, *125*, 16050–16057. [CrossRef] [PubMed]

27. Zhuang, Z.B.; Peng, Q.; Zhang, B.; Li, Y.D. Controllable synthesis of Cu2S nanocrystals and their assembly into a superlattice. *J. Am. Chem. Soc.* **2008**, *130*, 10482–10483. [CrossRef] [PubMed]

28. Liu, Z.P.; Xu, D.; Liang, J.B.; Shen, J.M.; Zhang, S.Y.; Qian, Y.T. Growth of Cu2S ultrathin nanowires in a binary surfactant solvent. *J. Phys. Chem.* **2005**, *109*, 10699–10704. [CrossRef] [PubMed]

29. Yin, C.Y.; Minakshi, M.; Ralph, D.E.; Jiang, Z.T.; Xie, Z.H.; Guo, H. Hydrothermal synthesis of cubic α-fe2O3, microparticles using glycine: Surface characterization, reaction mechanism and electrochemical activity. *J. Alloys Compd.* **2011**, *509*, 9821–9825. [CrossRef]

30. Zhang, Y.Q.; Zhang, B.P.; Ge, Z.H.; Zhu, L.F.; Li, Y. Preparation by solvothermal synthesis, growth mechanism, and photocatalytic performance of CuS nanopowders. *Eur. J. Inorg. Chem.* **2014**, *2014*, 2368–2375. [CrossRef]

31. Barmi, M.J.; Minakshi, M. Tuning the redox properties of the nanostructured CoMoO4 electrode: Effects of surfactant content and synthesis temperature. *ChemPlusChem* **2016**, *81*, 964–977. [CrossRef]

32. Penki, T.R.; Shivakumara, S.; Minakshi, M.; Munichandraiah, N. Porous Flower-like α-Fe2O3 Nanostructure: A High Performance Anode Material for Lithium-ion Batteries. *Electrochim. Acta* **2015**, *167*, 330–339. [CrossRef]

33. He, Y.; Day, T.; Zhang, T.S.; Liu, H.L.; Shi, X.; Chen, L.D.; Snyder, G.J. High Thermoelectric Performance in Non-Toxic Earth-Abundant Copper Sulfide. *Adv. Mater.* **2014**, *26*, 3974–3978. [CrossRef] [PubMed]

34. Ge, Z.H.; Liu, X.Y.; Feng, D.; Lin, J.Y.; He, J.Q. High-Performance Thermoelectricity in Nanostructured Earth-Abundant Copper Sulfides Bulk Materials. *Adv. Energy Mater.* **2016**, *6*, 1600607. [CrossRef]

35. Li, B.; Huang, L.; Zhao, G.Y.; Wei, Z.M.; Dong, H.L.; Hu, W.P.; Wang, L.W.; Li, J.B. Large-Size 2D β-Cu2S Nanosheets with Giant Phase Transition Temperature Lowering (120 K) Synthesized by a Novel Method of Super-Cooling Chemical-Vapor-Deposition. *Adv. Mater.* **2016**, *28*, 8271–8276. [CrossRef] [PubMed]

36. Okamoto, K.; Kawai, S. Electrical conduction and phase transition of copper sulfides. *Jpn. J. Appl. Phys.* **1973**, *12*, 1130. [CrossRef]

37. Ge, Z.H.; Zhao, L.D.; Wu, D.; Zhang, B.P.; Li, J.F.; He, J.Q. Low cost, abundant binary sulfides as promising thermoelectric. *Mater. Today* **2016**, *19*, 227–239. [CrossRef]

© 2017 by the authors. Licensee MDPI, Basel, Switzerland. This article is an open access article distributed under the terms and conditions of the Creative Commons Attribution (CC BY) license (http://creativecommons.org/licenses/by/4.0/).

crystals

MDPI

Article

Enhanced Thermoelectric Properties of Graphene/Cu$_2$SnSe$_3$ Composites

Degang Zhao *, Xuezhen Wang and Di Wu

School of Materials Science and Engineering, University of Jinan, Jinan 250022, China;
ujn_wangxz@yeah.net (X.W.); crystal4885@sina.com (D.W.)
* Correspondence: mse_zhaodg@ujn.edu.cn; Tel.: +86-531-8276-7561

Academic Editor: George S. Nolas
Received: 28 January 2017; Accepted: 27 February 2017; Published: 28 February 2017

Abstract: Cu$_2$SnSe$_3$ material is regarded as a potential thermoelectric material due to its relatively high carrier mobility and low thermal conductivity. In this study, graphene was introduced into the Cu$_2$SnSe$_3$ powder by ball milling, and the bulk graphene/Cu$_2$SnSe$_3$ thermoelectric composites were prepared by spark plasma sintering. The graphene nanosheets distributed uniformly in the Cu$_2$SnSe$_3$ matrix. Meanwhile, some graphene nanosheets tended to form thick aggregations, and the average length of these aggregations was about 3 μm. With the fraction of graphene increasing, the electrical conductivity of graphene/Cu$_2$SnSe$_3$ samples increased greatly while the Seebeck coefficient was decreased. The introduction of graphene nanosheets can reduce the thermal conductivity effectively resulting from the phonon scattering by the graphene interface. When the content of graphene exceeds a certain value, the thermal conductivity of graphene/Cu$_2$SnSe$_3$ composites starts to increase. The achieved highest figure of merit (*ZT*) for 0.25 vol % graphene/Cu$_2$SnSe$_3$ composite was 0.44 at 700 K.

Keywords: thermoelectric; composites; ternary diamond-like semiconductor; graphene

1. Introduction

Due to the dilemma between energy crisis and environmental stewardship, developing renewable energy technologies has attracted considerable research interest in the past decade. Thermoelectric materials, which can directly convert heat energy into electrical energy and vice versa, show great promise in the application of solid-state cooling, waste heat recovery, and power generation. The conversion efficiency of thermoelectric material is governed by the dimensionless figure of merit, $ZT = \sigma\alpha^2 T/\kappa$, where σ, α, T, and κ are the electrical conductivity, Seebeck coefficient, absolute temperature and thermal conductivity, respectively. The total thermal conductivity is composed of carrier thermal conductivity (κ_c) and lattice thermal conductivity (κ_l). Therefore, thermoelectric materials with good performance should have a large α and σ and low κ. As the fundamental material parameters (α, σ, and κ_c) are interrelated and conflicting via carrier concentration in bulk thermoelectric materials, it is a longstanding challenge to largely improve the overall ZT [1–4]. Therefore, concepts or strategies that can decouple these parameters to simultaneously optimize the electron and phonon transport are highly encouraging and imperative for the thermoelectric community. Specifically, band engineering and nanostructuring have been demonstrated as effective extrinsic approaches to separately enhance the power factor ($PF = \alpha^2\sigma$) and reduce the κ_l, respectively.

Several classes of thermoelectric materials, such as skutterudite [5,6], tellurides [7–10], half-Heuslers [11,12], and silicides [13,14], have been modified to reach high ZT value. Recently, ternary diamond-like semiconductor of Cu$_2$SnSe$_3$ has emerged as a new potential thermoelectric material due to its relatively high carrier mobility and quite low thermal conductivity. Since the Cu–Se bond network in the Cu$_2$SnSe$_3$ structure forms an electrically conductive framework and Sn orbitals contribute little to the carrier transport, the electrical conductivity of Cu$_2$SnSe$_3$ is allowed to be tuned to optimize the thermoelectric property by partial substitution of the Sn site. Some valuable work has been done on Cu_2SnSe_3 compound by doping, substitution, or solid solution [15,16]. The In-doped $Cu_2In_xSn_{1-x}Se_3$ was studied by Chen et al. and a maximum *ZT* of around 1.2 was obtained at 850 K for $x = 0.1$ [17]. Similarly, gallium doping was found to be an effective way to increase the *ZT* in Ga-doped Cu_2SnSe_3 compounds by Shi et al., and the maximum *ZT* increased to 0.43 at 700 K [18]. Moreover, isoelectronic alloying with Ge at the Sn site was confirmed to be effective in enhancing the *ZT* value by Morelli et al. [19]. Besides substitution, the introduction of a nanostructure phase into the matrix is also an attractive approach to enhance the dimensionless figure of merit of thermoelectric materials. So far, there are few studies about nanostructured Cu_2SnSe_3 matrix composites due to the unapparent enhancement of *ZT* resulting from the second nanostructured phase. Although a remarkable decrease in the lattice conductivity can be achieved by phonon scattering at nanophase/matrix interfaces, the electrical properties of thermoelectric composites also decrease, leading to a marginal change of the overall *ZT* value. Moreover, the selection of nanophase and the control of the microstructure of thermoelectric composites are also important for the enhancement of *ZT* value [20–22].

Graphene has high electrical and thermal properties due to its unique 2D structure. The carrier mobility, electrical conductivity, and thermal conductivity of graphene is 2×10^5 $cm^2 \cdot V^{-1} \cdot s^{-1}$, 1×10^6 S/m, and 5×10^3 $Wm^{-1} \cdot K^{-1}$ at room temperature, respectively. Meanwhile, the carrier of graphene with zero bandgap can continuously vary from electron to hole, which can benefit the electrical transport in the p–n interfacial region. Wang et al. even confirmed that the introduction of 0.2 vol % graphene enhanced the *ZT* value of Bi_2Te_3 material [23]. Kim et al. confirmed that the peak *ZT* value for the 0.05 wt % graphene/$Bi_2Te_{2.7}Se_{0.3}$ composite increased to 0.8 at 400 K, which is 23% larger than that of the pristine sample [24]. Chen et al. also showed that an improved *ZT* value of 0.4 in graphene/$CuInTe_2$ composites was obtained due to a lower κ_l [25]. In this contribution, it is highly possible that incorporating graphene nanosheets into Cu_2SnSe_3 material will also lead to reduced κ_l, which perhaps will further improve the thermoelectric properties of graphene/Cu_2SnSe_3 composites.

In the present work, graphene nanosheets were incorporated into the Cu_2SnSe_3 matrix by ball-milling method, and the graphene/Cu_2SnSe_3 thermoelectric composites were fabricated by spark plasma sintering (SPS). The transport properties of graphene/Cu_2SnSe_3 composites were studied with the aim of enhancing thermoelectric performance of Cu_2SnSe_3.

2. Experimental Procedures

Cu_2SnSe_3 was synthesized by the reacting stoichiometric copper (powder, 99.96%), tin (powder, 99.999%), and selenium (shot, 99.999%) in evacuated fused-silica ampoules at 1173 K for 12 h, then slowly cooling the melt down to 873 K for 24 h, followed by annealing at this temperature for 2 days. Finally, the obtained ingots were reground into fine powder. Commercially available graphene powder (single layer, average diameter: 2 μm, thickness: 0.8 nm; XFNANO, Nanjing, China) was chosen as the second nanophase, just as shown in Figure 1. The graphene powder was incorporated into the Cu_2SnSe_3 powder at volume fractions of 0.25, 0.50, 0.75, and 1.0 vol %, respectively. Then, the graphene-added Cu_2SnSe_3 powders were mechanically milled with a planetary ball-milling machine. The ball-to-powder ratio was 5:1, and the ball-milling process was carried out in Ar atmosphere for 240 min at 150 rpm. The SPS process was used to consolidate the as-milled powders at 860 K for 8 min in a vacuum of 0.1 Pa under a pressure of 50 MPa.

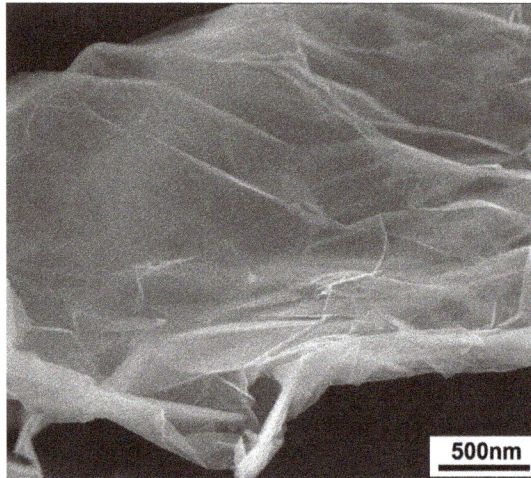

Figure 1. SEM image of single-layer graphene with an average diameter of 2 μm.

The constituent phases of the samples were characterized by X-ray diffractometry on a Rigaku Rint2000 powder diffractometer equipped with Cu K$_\alpha$ radiation. The microstructure of all graphene/Cu$_2$SnSe$_3$ samples was observed using field-emission scanning electron microscopy (FESEM) and high-resolution transmission electron microscopy (HRTEM, JEM2100F, JEOL, Tokyo, Japan). The thermal diffusivity (λ) of all samples was measured on the disk-shaped specimen by laser flash technique using a Netzsch LFA427 (Netzsch, Berlin, German) setup in a flowing Ar atmosphere with temperature ranging from 300 to 700 K. The thermal conductivity was then calculated as $\kappa = d\lambda C_p$, where d is the density measured by Archimedes method, and C_p is the Dulong–Petit approximation for the specific heat capacity. A bar-shaped specimen of $2 \times 2 \times 10$ mm^3 was cut with a diamond saw from the sample for the measurement of electrical transport properties. Both electrical conductivity and Seebeck coefficient were determined simultaneously using ZEM-3 equipment (ULVAC-RIKO, Tokyo, Japan) with temperature ranging from 300 to 700 K in Ar atmosphere. The Hall coefficients (R_H) were measured by van der Pauw's method in a vacuum of 0.1 Pa under a magnetic field of 2 T. The carrier mobility (μ$_H$) and carrier concentration (p) were calculated through the formulae of $\mu_H = R_H\sigma$ and $p_H = 1/(R_He)$ based on the assumption of single-band model, where e is the electron charge. The experimental uncertainty on the electrical conductivity, Seebeck coefficient, thermal conductivity, and Hall coefficient are estimated to be 5%, 5%, 8%, and 4%, respectively.

3. Results and Discussion

3.1. Phase Analysis and Microstructure

Figure 2 displays the SEM image of the 1.0 vol % graphene-added Cu$_2$SnSe$_3$ powder after ball milling for 240 min at 150 rpm. It can be observed that the average diameter of graphene nanosheets in the mixed Cu$_2$SnSe$_3$ powder was about 1 μm. Figure 3 is the X-ray diffraction patterns of sintered graphene/Cu$_2$SnSe$_3$ composites. The diffraction peaks in Figure 3 are identified as JCPDS card 65-4145 (cubic Cu$_2$SnSe$_3$). No diffraction peak of graphene is found in the XRD results as the fraction of graphene in the composites is very low. All graphene/Cu$_2$SnSe$_3$ composites show the same XRD patterns as the pristine Cu$_2$SnSe$_3$.

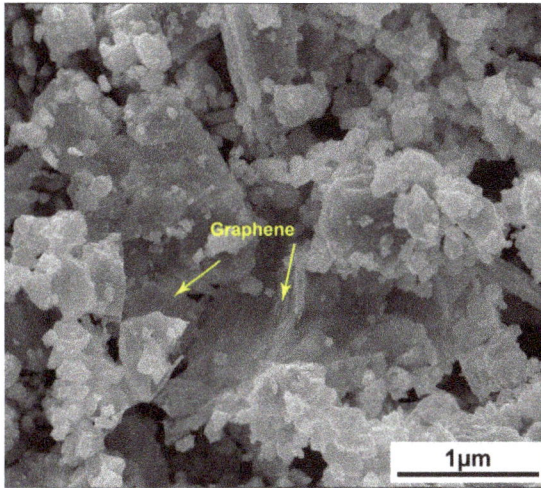

Figure 2. SEM image of the 1.0 vol % graphene/Cu$_2$SnSe$_3$ powder after ball milling.

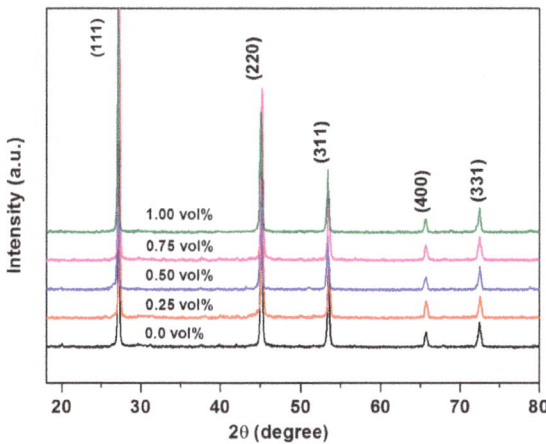

Figure 3. XRD patterns of sintered graphene/Cu$_2$SnSe$_3$ samples.

The microstructure of the sintered pristine Cu$_2$SnSe$_3$ and 0.75 vol % graphene/Cu$_2$SnSe$_3$ sample is illustrated in Figure 4a,b, respectively. It is evident that the graphene nanosheets distributed uniformly in the Cu$_2$SnSe$_3$ matrix. Meanwhile, some graphene nanosheets tended to form thick aggregations and the average length of aggregations was about 3 μm. A similar phenomenon was also observed by Zhao et al. in the graphene/CoSb$_3$ nanocomposite [26]. The results of energy dispersive X-ray spectroscopy (EDS) for graphene/Cu$_2$SnSe$_3$ sample identify that the matrix consisted of 33.17 atom % copper, 16.79 atom % tin, and 50.04 atom % selenium, indicating the Cu$_2$SnSe$_3$ phase, just as shown in Figure 5. The black phase in Figure 5a only contains the C element, corresponding to the graphene phase. It can be observed from HRTEM in Figure 6 that most of the graphene nanosheets is of the multilayered form (<10 layers), which is consistent with SEM results. The fringe spacing of 0.81 nm in the lattice image corresponds to the interplanar distance of the (111) plane of Cu$_2$SnSe$_3$. Figure 7 shows the FESEM image of the fractured surface of the sintered graphene/Cu$_2$SnSe$_3$ sample. The graphene nanosheets are homogeneously embedded in the Cu$_2$SnSe$_3$ matrix. According to the classic band theory [27,28], nanostructures distributed in the material can result in strain fields, then lead to a change in the energy-band structure of thermoelectric material. At the same time, nanophases can greatly influence the phonon and electronic transport of thermoelectric materials.

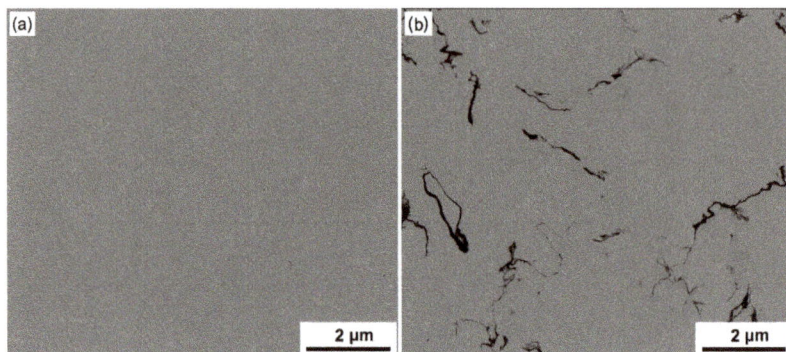

Figure 4. SEM image of the sintered (**a**) Cu_2SnSe_3; (**b**) 0.75 vol % graphene/Cu_2SnSe_3 sample.

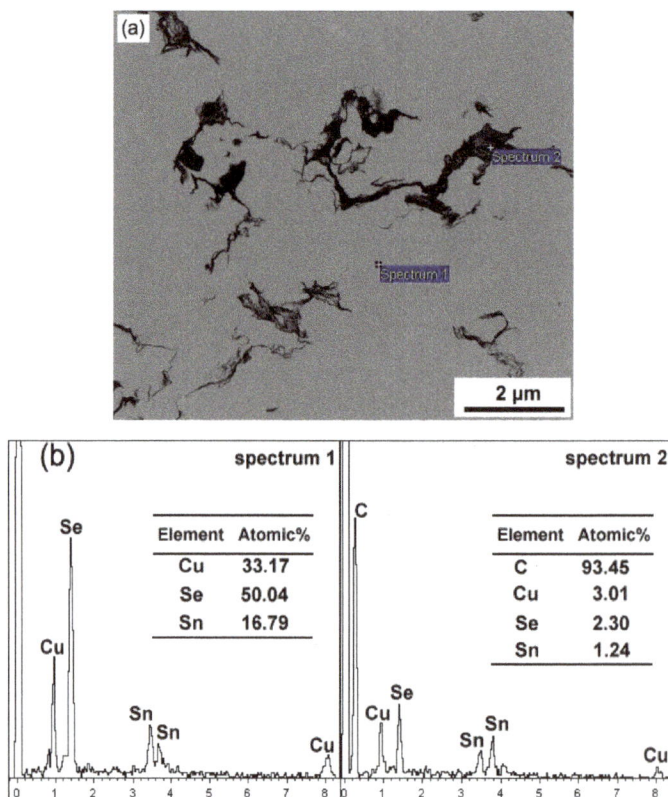

Figure 5. (**a**) SEM image of the sintered 1.0% graphene/Cu_2SnSe_3 composite; (**b**) energy dispersive X-ray spectroscopy (EDS) analysis.

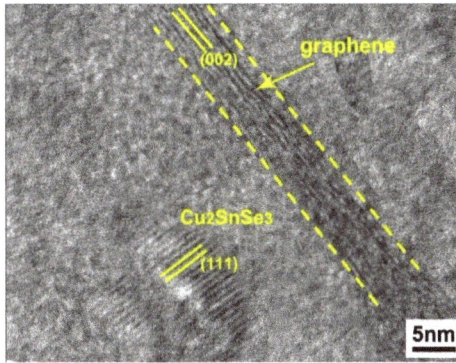

Figure 6. High-resolution TEM (HRTEM) image of graphene nanosheets in the graphene/Cu_2SnSe_3 sample.

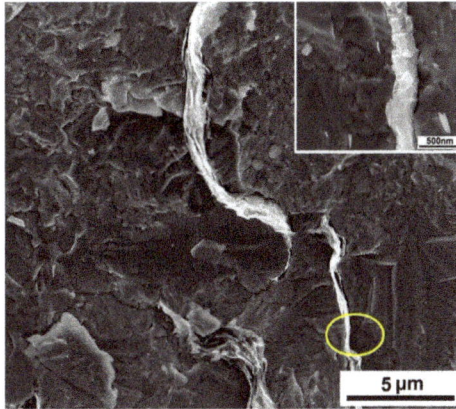

Figure 7. Field-emission SEM (FESEM) image of fractured surface of the sintered graphene/Cu_2SnSe_3 sample.

3.2. Electrical Properties

Figure 8 presents the σ of graphene/Cu_2SnSe_3 composites as a function of temperature. It can be observed that the σ of the pristine Cu_2SnSe_3 sample declines steeply with the temperature increasing across the overall temperature range, showing a typical heavily doped degenerate semiconducting behavior. It is noteworthy that graphene/Cu_2SnSe_3 samples show an obvious increased σ compared with pristine Cu_2SnSe_3 due to the introduction of conductivity graphene nanosheets. In addition, the σ of graphene/Cu_2SnSe_3 samples increases with the increasing fraction of graphene. The σ of 1.0 vol % graphene/Cu_2SnSe_3 sample at room temperature is about 350 $\Omega^{-1}\cdot cm^{-1}$, which is about 3 times the value of the pristine Cu_2SnSe_3. Even at the high-temperature region, the σ of graphene/Cu_2SnSe_3 sample still retains a high value. The σ of 0.25 vol % graphene/Cu_2SnSe_3 sample is around 124 $\Omega^{-1}\cdot cm^{-1}$ at 700 K. The enhancement in σ for graphene/Cu_2SnSe_3 samples may be ascribed to either an increase of carrier concentration (p), or the increment in carrier mobility (μ_H), or both. Table 1 lists some physical and structural parameters of the graphene/Cu_2SnSe_3 composites at room temperature. As shown in Table 1, the carrier concentration of graphene/Cu_2SnSe_3 composites is higher than that of pristine Cu_2SnSe_3. The carrier mobility increases from 21.2 $cm^2/V\cdot s$ for Cu_2SnSe_3 to 34.3 $cm^2/V\cdot s$ for the 1.0 vol % graphene/Cu_2SnSe_3 sample. Therefore, it can be concluded that incorporating graphene nanosheets into a Cu_2SnSe_3 matrix can improve the electrical conductivity, which is attributed to the increment in both carrier concentration and mobility. This is reasonable because the multilayered graphene is *p*-type thermoelectric material, and the graphene itself can afford the charged carrier [29]. In addition, the graphene has a relatively high mobility, which is beneficial to increase the carrier mobility of graphene/Cu₂SnSe₃ composites.

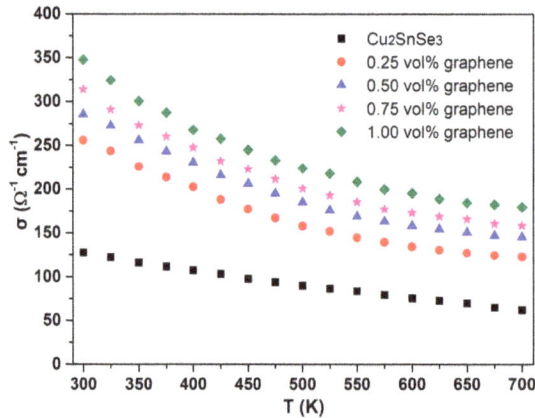

Figure 8. Electrical conductivity of graphene/Cu₂SnSe₃ samples as a function of temperature.

Table 1. Chemical composition and some physical and structural parameters of graphene/Cu₂SnSe₃ composites at room temperature.

x (vol %)	Relative Density	σ ($\Omega^{-1}\cdot cm^{-1}$)	p (10^{19} cm^{-3})	μ_H (cm²/V·s)	α (μV/K)	κ_l (W·m^{-1}·K^{-1})	m^* (m_0)
0	98.7%	127	3.74	21.2	131	2.65	2.6
0.25	98.1%	255	5.43	29.3	99.8	2.36	2.8
0.50	98.0%	285	5.86	30.4	92.0	2.59	2.9
0.75	97.8%	313	6.13	31.9	78.7	2.92	2.7
1.00	97.5%	448	8.16	34.3	69.9	3.24	3.1

x: volume fraction; σ: electrical conductivity; p: charge carrier concentration; μ_H: carrier mobility; α: Seebeck coefficient; κ_l: thermal conductivity; m^*: density of states effective mass

Figure 9 demonstrates the α of graphene/Cu₂SnSe₃ samples as a function of temperature. It can be seen that the α of all graphene/Cu₂SnSe₃ samples across the whole temperature range was positive, indicating the major charge carriers in the samples are holes. Moreover, the α of all graphene/Cu₂SnSe₃ composites and pristine Cu₂SnSe₃ samples increases approximately linearly with increasing temperature. For example, the α of pristine Cu₂SnSe₃ increases from 130 µV/K to 255 µV/K in the temperature range of 300–700 K. At the same time, the introduction of graphene nanosheets decreased the Seebeck coefficients of Cu₂SnSe₃ samples evidently. Compared with the α of pristine Cu₂SnSe₃ sample, the α of graphene/Cu₂SnSe₃ samples decreases with the increasing fraction of graphene. At room temperature, the α decreases from 130 µV/K for Cu₂SnSe₃ matrix to 70 µV/K for the 1.0% graphene/Cu₂SnSe₃ composite. The decrease of α for graphene/Cu₂SnSe₃ composites can be explained by the equation

$$\alpha = \pm\frac{k_B}{e}\left[2 + \ln\frac{2(2\pi m^* k_B T)^{\frac{3}{2}}}{h^3 p}\right] \qquad (1)$$

where k_B, m^*, h, and p are Boltzmann constant, density of states effective mass, Planck's constant, and charge carrier concentration, respectively. The introduction of graphene nanosheets leads to the improved carrier density. Herein, according to the equation, the α is reduced.

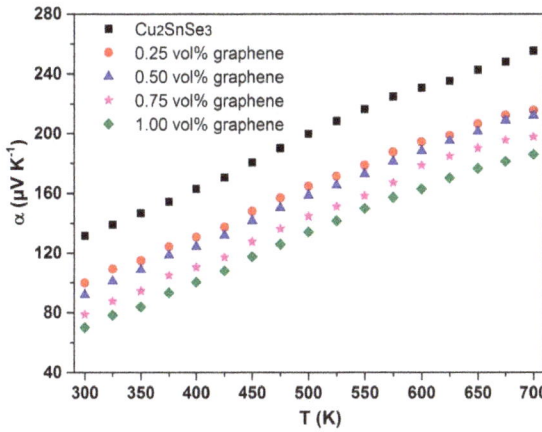

Figure 9. Seebeck coefficient (α) of graphene/Cu$_2$SnSe$_3$ samples as a function of temperature.

The μ_H of graphene/Cu$_2$SnSe$_3$ composites as a function of temperature is displayed in Figure 10. The μ_H of graphene/Cu$_2$SnSe$_3$ composites increases with the increasing fraction of graphene. Moreover, the μ_H of graphene/Cu$_2$SnSe$_3$ samples in this study is between 20 and 35 cm$^2\cdot$V$^{-1}\cdot$s^{-1} at room temperature, which is close with that of CoSb$_3$ [30,31]. This may be attributed to the similar carrier effective mass (m^*) of Cu$_2$SnSe$_3$ and skutterudite compounds. The m^* can be calculated by the following equations based on single parabolic band model.

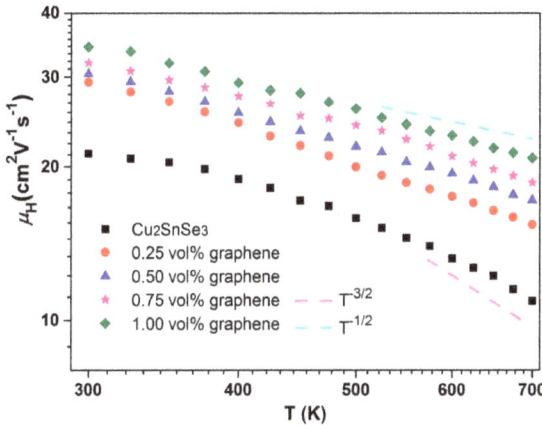

Figure 10. Carrier mobility (μ_H) of graphene/Cu$_2$SnSe$_3$ samples as a function of temperature.

Table 1 lists the evaluated equivalent m^* of graphene/Cu$_2$SnSe$_3$ samples at room temperature. Meanwhile, it can also be observed in Figure 10 that the μ_H of pristine Cu$_2$SnSe$_3$ shows a temperature dependence of $T^{-3/2}$ above 520 K, suggesting that the dominant scattering mechanism is phonon scattering in the temperature range from 520 K to 700 K. Below 520 K, the μ_H of pristine Cu$_2$SnSe$_3$ proportional to $T^{-3/2}$ is weak, and the relationship of μ_H as function of temperature dependence of $T^{-0.5}$ can be seen, showing that alloy scattering is the dominate mechanism in this temperature range. However, the μ_H of graphene/Cu$_2$SnSe$_3$ samples deviates from the $T^{-1.5}$ or $T^{-0.5}$ dependence across the entire temperature range, indicating the dominative mechanism is mixed scattering in these composites.

3.3. Thermal Conductivity

The κ and κ_l for graphene/Cu_2SnSe_3 samples as function of temperature is shown in Figure 11. The κ_l is obtained by directly subtracting the carrier thermal conductivity κ_c from the total thermal conductivity; κ_c can be calculated according to the Wiedemann–Franz law, $\kappa_c = L_0\sigma T$, where the Lorenz constant L_0 is taken as 2.45×10^{-8} V^2/K^2. The κ for all composites decreases with the increasing temperature. With the fraction of graphene increasing, the κ of graphene/Cu_2SnSe_3 composites firstly declines then starts to increase. The achieved κ of 0.25% graphene/Cu_2SnSe_3 sample at room temperature is 2.5 W/m·K, which is a 12% reduction from that of pristine Cu_2SnSe_3. On the contrary, the κ of the 1.0% graphene/Cu_2SnSe_3 sample at room temperature increases to 3.45 W/m·K. The κ_l of graphene/Cu_2SnSe_3 samples demonstrates similar changes compared to that of pure Cu_2SnSe_3. The lowest κ_l of 0.25% graphene/Cu_2SnSe_3 samples is 0.78 W/m·K, which is 22% lower than that of the pristine Cu_2SnSe_3 sample. As is known to all, nanostructuring will reduce the κ_l of material as the long-wavelength phonon scattering at grain boundaries was suppressed. Because graphene itself has high lattice thermal conductivity and large specific surface area, an opposite effect of graphene nanosheets on the κ of Cu_2SnSe_3 can be allowed. On one side, the addition of second phase with high κ may increase the total thermal conductivity of composite. On the other side, large specific surface area suggests more newly formed interfaces between the matrix and second phase, which are expected to scatter phonons to depress the κ_l. For 0.25% graphene/Cu_2SnSe_3, graphene nanosheets are homogeneously dispersed in the Cu_2SnSe_3 matrix, which means the dominative factor should be the influence of interface scattering. By comparison, when the content of graphene exceeds a certain value, the graphene in the composites tends to aggregate into thick flakes in the Cu_2SnSe_3 matrix, as mentioned above. Therefore, the interfacial increment due to the incorporation of graphene should not be significant. This can explain the change in κ_l of graphene/Cu_2SnSe_3 composites. The results also confirm that the κ_l of Cu_2SnSe_3 can be effectively reduced by introducing graphene nanosheets. The obtained minimum κ_l in the present work is 0.78 W/m·K at 700 K for the 1.0% graphene/Cu_2SnSe_3 sample. According to the basic kinetic theory, when the phonon mean free path is equal to the shortest interatomic distance, the lattice thermal conductivity can achieve the minimal value κ_{lmin} [32]. The κ_{lmin} can be calculated according to the formula $\kappa_l = 1/3v_mC_v\cdot l$, where v_m, C_v, and l are the mean sound velocity, the isochoric specific heat of the system using Dulong and Petit value, and the mean free path of phonon, respectively. The v_m is taken as the constant 2.3×10^3 m/s [33]. It is assumed that the minimum mean free path of phonon l is the interatomic distance (0.238 nm) of the Cu_2SnSe_3 structure, and the achieved κ_{lmin} is 0.52 W·m^{-1}·K^{-1}, just as illustrated by the brown dashed line in Figure 11b. By controlling the content of graphene nanosheets and microstructure of composites, the κ_l of graphene/Cu_2SnSe_3 composites may approach the κ_{lmin} of Cu_2SnSe_3 in the high-temperature region. Further optimization will be studied in further work.

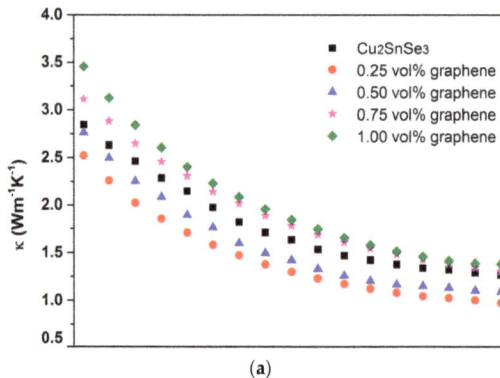

(a)

Figure 11. *Cont.*

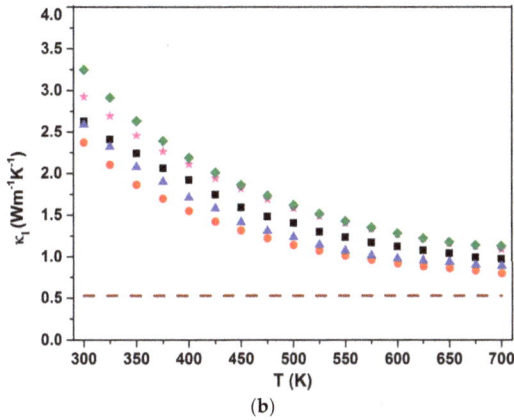

(b)

Figure 11. (a) Total thermal conductivity (κ) and (b) lattice thermal conductivity (κ_l) of graphene/Cu$_2$SnSe$_3$ samples as a function of temperature.

3.4. Figure of Merit

Figure 12 shows the ZT value of graphene/Cu$_2$SnSe$_3$ samples as a function of temperature. Like other related Cu-based ternary chalcogenide compounds with diamond-like structure [17,18], the ZT value of graphene/Cu$_2$SnSe$_3$ samples increases with increasing temperature. Compared with the figure of merit of pristine Cu$_2$SnSe$_3$, the ZT of graphene/Cu$_2$SnSe$_3$ samples is obviously improved. The 0.25% graphene/Cu$_2$SnSe$_3$ composite has the maximal ZT value of 0.44 at 700 K, 45% higher than that of pristine Cu$_2$SnSe$_3$. If the graphene/Cu$_2$SnSe$_3$ samples were coated by a coating film and the measured temperature increased to 850 K, the ZT value is capable of reaching 1.0–1.2. The enhancement of ZT for graphene/Cu$_2$SnSe$_3$ composites is basically ascribed to the depressed κ_l and the increased σ. The incorporation of graphene nanosheets into the Cu$_2$SnSe$_3$ could enhance the thermoelectric properties. Therefore, if we choose the material with optimized carrier concentration and mobility as the thermoelectric matrix, the thermoelectric composite with a higher ZT value could be achieved.

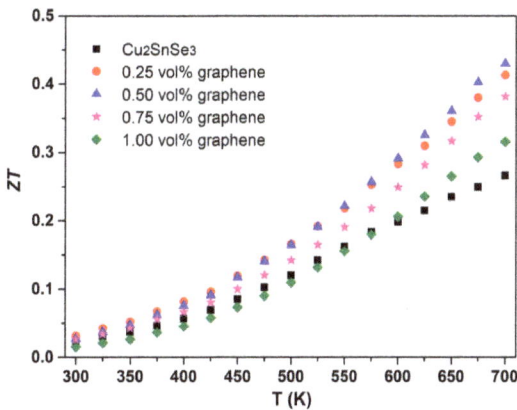

Figure 12. The dimensionless figure of merit of (ZT) of graphene/Cu$_2$SnSe$_3$ samples as a function of temperature.

4. Conclusions

The graphene nanosheets were introduced into the Cu_2SnSe_3 matrix by ball milling and the graphene/Cu_2SnSe_3 composite was fabricated by spark plasma sintering. The graphene nanosheets distributed uniformly in the Cu_2SnSe_3 matrix. Meanwhile, some graphene nanosheets tended to form thick aggregations and the average length of aggregations was about 3 μm. With the increasing content of graphene, the electrical conductivity of graphene/Cu_2SnSe_3 samples greatly increased, while the Seebeck coefficient was decreased. The introduction of graphene nanosheets reduced the thermal conductivity, effectively resulting from the phonon scattering by the graphene interface. When the fraction of graphene exceeds a certain value, the thermal conductivity of graphene/Cu_2SnSe_3 composites starts to increase. The maximum figure of merit ZT for 0.25 vol % graphene/Cu_2SnSe_3 was 0.44 at 700 K.

Acknowledgments: This work is financially supported by National Natural Science Foundations of China (Grants Nos. 51471076, 51202088).

Author Contributions: All authors participated in the research, analysis and edition of the manuscript. Degang Zhao designed the experiments and Xuezhen Wang fabricated the samples. All authors contributed to the characterization and data analysis. Degang Zhao wrote the paper.

Conflicts of Interest: The authors declare no conflicts of interest.

References

1. Sootsman, J.R.; Chung, D.Y.; Kanatzidis, M.G. New and old concepts in thermoelectric materials. *Angew. Chem. Int. Ed.* **2009**, *48*, 8616–8639. [CrossRef] [PubMed]
2. Shi, X.; Chen, L.D.; Uher, C. Recent advance in high-performance bulk thermoelectric materials. *Int. Mater. Rev.* **2016**, *61*, 379–415. [CrossRef]
3. Gayner, C.; Kar, K.K. Recent advances in thermoelectric materials. *Prog. Mater. Sci.* **2016**, *83*, 330–382. [CrossRef]
4. Zhao, L.D.; Kanatzidis, M.G. An overview of advanced thermoelectric materials. *J. Materiomics* **2016**, *2*, 101–103. [CrossRef]
5. Zhou, X.Y.; Wang, G.W.; Guo, L.J.; Chi, H.; Wang, G.Y.; Zhang, Q.F.; Chen, C.Q.; Travis, T.; Jeff, S.; Dravid, V.P. Hierarchically structured TiO_2 for Ba-filled skutterudite with enhanced thermoelectric performance. *J. Mater. Chem. A* **2014**, *2*, 20629–20635. [CrossRef]
6. Li, G.D.; Baijaj, S.; Aydemir, U.; Hao, S.Q.; Xiao, H.; Goddard, W.A.; Zhai, P.C.; Zhang, Q.J.; Snyder, G.J. P-type Co interstitial defects in thermoelectric skutterudite $CoSb_3$ due to the breakage of Sb_4 rings. *Chem. Mater.* **2016**, *28*, 2172–2179. [CrossRef]
7. Gelbstein, Y.; Dashevsky, Z.; Dariel, M.P. In-doped $Pb_{0.5}Sn_{0.5}Te$ p-type samples prepared by powder metallurgical processing for thermoelectric applications. *Physica B* **2007**, *396*, 16–21. [CrossRef]
8. Gelbstein, Y.; Davidow, J. Highly efficient functional $Ge_xPb_{1-x}Te$ based thermoelectric alloys. *Phys. Chem. Chem. Phys.* **2014**, *16*, 20120–20126. [CrossRef] [PubMed]
9. Gelbstein, Y. Phase morphology effects on the thermoelectric properties of $Pb_{0.25}Sn_{0.25}Ge_{0.5}Te$. *Acta Mater.* **2013**, *61*, 1499–1507. [CrossRef]
10. Dado, B.; Gelbstein, Y.; Mogilianasky, D.; Ezersky, V.; Dariel, M.P. Structural evolution following spinodal decomposition of the pseudoternary compound $(Pb_{0.3}Sn_{0.1}Ge_{0.6})Te$. *J. Electron. Mater.* **2010**, *39*, 2165–2171. [CrossRef]
11. Kirievsky, K.; Shlimovich, M.; Fuks, D.; Gelbstein, Y. An ab-initio study of the thermoelectric enhancement potential in nano-grained TiNiSn. *Phys. Chem. Chem. Phys.* **2014**, *16*, 20023–20029. [CrossRef] [PubMed]
12. Kirievsky, K.; Gelbstein, Y.; Fuks, D. Phase separation and antisite defects possibilities for enhancement the thermoelectric efficiency in TiNiSn half-Heusler alloys. *J. Solid State Chem.* **2013**, *203*, 247–254. [CrossRef]
13. Sadia, Y.; Dinnerman, L.; Gelbstein, Y. Mechanical Alloying and Spark Plasma Sintering of Higher Manganese Silicides for Thermoelectric Application. *J. Electron. Mater.* **2013**, *42*, 1926–1931. [CrossRef]

14. Gelbstein, Y.; Tunbridge, J.; Dixon, R.; Reece, M.J.; Ning, H.P.; Gilchrist, R.; Summers, R.; Agote, I.; Lagos, M.A.; Simpson, K.; et al. Physical, mechanical and structural properties of highly efficient nanostructured n- and p-silicides for practical thermoelectric applications. *J. Electron. Mater.* **2014**, *43*, 1703–1711. [CrossRef]

15. Liu, G.H.; Chen, K.X.; Li, J.T.; Li, Y.Y.; Zhou, M.; Li, L.F. Combustion synthesis of Cu_2SnSe_3 thermoelectric materials. *J. Eur. Ceram. Soc.* **2016**, *36*, 1407–1415. [CrossRef]

16. Lu, X.; Morelli, D.T. Thermoelectric properties of Mn doped Cu_2SnSe_3. *J. Electron. Mater.* **2012**, *41*, 1554–1558. [CrossRef]

17. Shi, X.Y.; Xi, L.L.; Fan, J.; Zhang, W.Q.; Chen, L.D. Cu–Se bond network and thermoelectric compounds with complex diamond like structure. *Chem. Mater.* **2010**, *22*, 6029–6031. [CrossRef]

18. Fan, J.; Liu, H.L.; Shi, X.Y.; Bai, S.Q.; Shi, X.; Chen, L.D. Investigation of thermoelectric properties of $Cu_2Ga_xSn_{1-x}Se_3$ diamond-like compounds by hot pressing and spark plasma sintering. *Acta Mater.* **2013**, *61*, 4297–4304. [CrossRef]

19. Skoug, E.J.; Cain, J.D.; Morelli, D.T. Thermoelectric properties of the Cu_2SnSe_3-Cu_2GeSe_3 solid solution. *J. Alloys Compd.* **2012**, *506*, 18–23. [CrossRef]

20. Bux, S.K.; Fleurial, J.P.; Kaner, R.B. Nanostructured materials for thermoelectric applications. *Chem. Commun.* **2011**, *46*, 8311–8324. [CrossRef] [PubMed]

21. Balaya, P. Size effects and nanostructured materials for energy applications. *Energy Environ. Sci.* **2008**, *1*, 645–654. [CrossRef]

22. Amatya, R.; Ram, R.J. Trend for thermoelectric materials and their earth abundance. *J. Electron. Mater.* **2012**, *41*, 1011–1558. [CrossRef]

23. Liang, B.B.; Song, Z.J.; Wang, M.H.; Wang, L.J.; Jiang, W. Fabrication and thermoelectric properties of graphene/Bi_2Te_3 composite materials. *J. Nanomater.* **2013**, *15*, 210767–210772.

24. Kim, J.; Lee, E.S.; Kim, J.Y.; Choi, S.M.; Lee, K.H.; Seo, W.S. Thermoelectric properties of unoxidized graphene/$Bi_2Te_{2.7}Se_{0.3}$ composites synthesized by exfoliation/re-assembly method. *Phys. Status Solidi (RRL)* **2014**, *8*, 1–5. [CrossRef]

25. Chen, H.J.; Yang, C.Y.; Liu, H.L.; Zhang, G.H.; Wan, D.Y.; Huang, F.Q. Thermoelectric properties of $CuInTe_2$/graphene composites. *CrystEngComm* **2013**, *15*, 6648–6651. [CrossRef]

26. Feng, B.; Xie, J.; Cao, G.S.; Zhu, T.J.; Zhao, X.B. Enhanced thermoelectric properties of p-type $CoSb_3$/graphene nanocomposite. *J. Mater. Chem. A* **2013**, *1*, 13111–13119. [CrossRef]

27. Faleev, S.V.; Leonard, F. Theory of enhancement of thermoelectric properties of materials with nanoinclusions. *Phys. Rev. B* **2008**, *77*, 214304. [CrossRef]

28. Zebarjadi, M.; Esfarjani, K.; Shakouri, A.; Zeng, G.; Lu, H.; Zide, J.; Gossard, A. Effect of nanoparticle scattering on thermoelectric power factor. *Appl. Phys. Lett.* **2009**, *94*, 202105. [CrossRef]

29. Dong, J.D.; Liu, W.; Li, H.; Su, X.L.; Tang, X.F.; Uher, C. In situ synthesis and thermoelectric properties of PbTe-graphene nanocomposites by utilizing a facile and novel wet chemical method. *J. Mater. Chem. A* **2013**, *1*, 12503–12511. [CrossRef]

30. Gharleghi, A.; Liu, Y.F.; Zhou, M.H.; He, J.; Tritt, T.M.; Liu, C.J. Enhancing the thermoelectric performance of nanosized $CoSb_3$ via short-range percolation of electrically conductive WTe_2 inclusions. *J. Mater. Chem. A* **2016**, *4*, 13874–13879. [CrossRef]

31. Ortiz, B.R.; Crawford, C.M.; Mckinney, R.W.; Parilla, P.A.; Toberer, E.S. Thermoelectric properties of bromine filled $CoSb_3$ skutterudite. *J. Mater. Chem. A* **2016**, *4*, 8444–8449. [CrossRef]

32. Morelli, D.T.; Jovovic, V.; Heremans, J.P. Intrinsically minimal thermal conductivity in cubic I-V-VI_2 semiconductors. *Phys. Rev. Lett.* **2008**, *101*, 035901. [CrossRef] [PubMed]

33. Cho, J.Y.; Shi, X.; Salvador, J.R.; Meisner, G.P. Thermoelectric properties and investigations of low thermal conductivity in Ga-doped Cu_2GeSe_3. *Phys. Rev. B* **2011**, *84*, 085207. [CrossRef]

© 2017 by the authors. Licensee MDPI, Basel, Switzerland. This article is an open access article distributed under the terms and conditions of the Creative Commons Attribution (CC BY) license (http://creativecommons.org/licenses/by/4.0/).

MDPI

St. Alban-Anlage 66

4052 Basel

Switzerland

Tel. +41 61 683 77 34

Fax +41 61 302 89 18

www.mdpi.com

Catalysts Editorial Office

E-mail: catalysts@mdpi.com

www.mdpi.com/journal/catalysts

www.ingramcontent.com/pod-product-compliance
Lightning Source LLC
Chambersburg PA
CBHW051913210326
41597CB00033B/6129